Molecular Therapeutics:

21st-century Medicine

Pamela Greenwell

University of Westminster, London, UK

Michelle McCulley

Peninsula Medical School,
Universities of Exeter and Plymouth, UK

John Wiley & Sons, Ltd

Other Wiley Editorial Offices

John Wiley & Sons Inc., 111 River Street, Hoboken, NJ 07030, USA

Jossey-Bass, 989 Market Street, San Francisco, CA 94103-1741, USA

Wiley-VCH Verlag GmbH, Boschstr. 12, D-69469 Weinheim, Germany

John Wiley & Sons Australia Ltd, 33 Park Road, Milton, Queensland 4064, Australia

John Wiley & Sons (Asia) Pte Ltd, 2 Clementi Loop #02-01, Jin Xing Distripark,
Singapore 129809

John Wiley & Sons Canada Ltd, 6045 Freemont Blvd, Mississauga, Ontario, L5R 4J3

Wiley also publishes its books in a variety of electronic formats. Some content that appears in
print may not be available in electronic books.

Anniversary Logo Design: Richard J Pacifico

Library of Congress Cataloging-in-Publication Data

Greenwell, Pamela.
    Molecular therapeutics : 21st century medicine / by Pamela Greenwell and Michelle
McCulley.
        p. ; cm.
    Includes index.
    ISBN 978-0-470-01916-0 (hardback : alk. paper) – ISBN 978-0-470-01917-7 (pbk. : alk.
paper)
    1. Gene therapy. 2. Molecular genetics. I. McCulley, Michelle. II. Title.
    [DNLM: 1. Gene Therapy – Popular Works. 2. Gene Therapy – ethics – Popular Works.
3. Gene Transfer Techniques – Popular Works. 4. Molecular Biology – Popular Works.
QZ 52 G816m 2007]
    RB155.8.G76 2007
    616′.042 – dc22   2007028135

British Library Cataloguing in Publication Data

A catalogue record for this book is available from the British Library

ISBN: 9-780-470-01916-0 (H/B)
        9-780-470-01917-7 (P/B)

Typeset in 10.5/13.5pt Sabon by SNP Best-set Typesetter Ltd., Hong Kong
Printed and bound in Great Britain by Antony Rowe Ltd, Chippenham, Wiltshire.
This book is printed on acid-free paper responsibly manufactured from sustainable forestry in
which at least two trees are planted for each one used for paper production.

Contents

Prologue

This book started its life in the last century, albeit 1990, as lecture notes for my undergraduate and postgraduate students studying Molecular Therapeutics. It has developed and evolved, spawning a distance learning package along the way. Despite calls from my students to publish the material as a textbook, I resisted. Then two people walked into my life: Nicky McGirr from John Wiley & Sons books and Michelle McCulley, a new lecturer, who managed to persuade me that, with their help, this book would and should be finished. And at last it is finished. I have enjoyed researching the area and, with Michelle's help and encouragement, writing the book. I hope it interests, informs and stimulates the reader. It is a fascinating area, rich in promise, full of ethical problems and some failures. I hope you enjoy it.

Molecular Therapeutics: 21st-century Medicine by Pamela Greenwell and Michelle McCulley.
© 2007 John Wiley & Sons, Ltd

1

Introduction

Molecular therapeutics is a new and developing science which involves genetics, recombinant DNA technology, biochemistry, protein production and purification, microbiology, molecular biology, immunology, pathobiology and biotechnology. It addresses the treatment of human beings with 'new drugs' and poses a range of ethical issues, particularly with respect to clinical trials, animal models, financial considerations and availability of treatment. Some therapies that have been developed in animal models, such as germ-line gene therapy and cloning, have been banned for use in humans and others are only available under strict licence.

The technology covers a wide range of disease therapies from protein supplementation to gene therapy and is applicable to microbial, inherited and acquired diseases. Some of the therapies to be discussed are currently being tested in cell culture. We need to address the validity of such studies in the light of the fact that cultured cells are not normal in as much as normal cells do not grow indefinitely in culture. Other therapies are being evaluated in animal models. We will need to discuss the ethics of the production of such models and the validity of results we may obtain. Other therapies to be described are still at the clinical trials stage and are not yet available for patient treatment. Such therapies are being evaluated by treating terminally ill patients who are unlikely to gain any benefit from the trials. Again, we must address the ethical and moral issues surrounding such practices.

Molecular Therapeutics: 21st-century Medicine by Pamela Greenwell and Michelle McCulley.
© 2007 John Wiley & Sons, Ltd

1.1 Microbial diseases

Diseases caused by bacteria, viruses and protozoans that affect mankind are common, particularly in the developing world. Indeed, during a year many of us will have suffered a virally induced infection, and worldwide the WHO estimates that 8000 people die of AIDS-related conditions every day, which equates to 3 million deaths per year. Every year there are 8.8 million new cases of tuberculosis (TB) reported, with 5500 deaths a day or a million deaths worldwide each year. Worldwide, there are about 300 million cases of acute cases of malaria reported each year.[1] However, many microbial diseases can be cured by the administration of antibiotics or prevented by vaccination. The WHO statistics for 2002 showed that 32% of people died from communicable diseases, 59% from non-communicable disease and the remaining 9% from injuries.[2] However, there are an increasing number of bacteria that are antibiotic-resistant, for example methicillin-resistant *Staphylococcus aureus* (MRSA) which can affect hospitalised immunocompromised patients and drug-resistant strains of *Mycobacterium tuberculosis* which are an increasing problem in patients infected with HIV. Drug-resistant strains of bacteria are thought to be caused by the overuse of antibiotics not only in human populations but also in farming where antibiotics are used to encourage fast growth of animals.

Viruses also pose a great threat to the well-being of humankind. For example, there are few effective treatments for human immunodeficiency virus (HIV), the causative agent of AIDS, human papilloma virus (HPV), which is implicated in cancer of the cervix, or hepatitis A, B and C which may cause long-term liver damage. It is also important to realise that we know much about the genome of these organisms but cannot easily formulate cures. All viral infections are difficult to treat effectively, some are potentially life-threatening and new viruses are emerging, for example Ebola virus.

Parasitic diseases are a particular problem in developing countries where the parasite burden leaves whole populations immunocompromised and open to opportunistic infection. Malaria for example affects hundreds of millions of people worldwide, but current treatment strategies are failing, new treatments are too expensive for use and no effective vaccines exist. There is therefore a need to develop new drugs, immunotherapy, gene therapy and vaccines to alleviate the problems. These therapies must be cheap enough to be used worldwide and stable enough to allow shipment and storage at room temperature.

In order to treat diseases effectively we need to understand the causative agent and the mode of infection. This may allow us to formulate prevention strategies that are simple and cost-effective. Simplistically, if we know the vector of a disease breeds in stagnant water then we may treat stagnant water with pesticide and kill the vector. Such an approach was successful in the prevention of malaria. In this case the stagnant pools were treated either with oil to decrease surface tension and essentially suffocate the larvae of the vector or with a pesticide to kill the larvae directly. Such schemes are effective until the vector develops resistance or until the political situation in the country becomes unstable, resulting in failure to treat the breeding grounds so that subsequently the problem re-emerges.

An understanding of the genetic and biochemical make-up of the organism allows us to formulate a range of strategies based on our knowledge. For example, analysis of the biochemistry of HIV showed that a protease was vital for cleaving a specific protein needed by the organism for survival. This protease was distinctly different from human proteases and therefore specific inhibitors could be designed to affect the viral enzyme.[3,4]

The host also plays a role in infection and its control. The host immune response will affect whether an infectious agent will be removed from the host or retained and allowed to multiply. The host genotype also plays a role. Again HIV is a good example where individuals with the rare genotype $ccr5^{-/-}$ lack a cell surface receptor vital for infectivity. Another example involves smallpox, which is known to carry the blood group A antigen. Individuals who are blood group O have natural antibodies to this structure, which will bind to the virus and elicit destruction via the immune response. Blood group A individuals lack this first line of defence since they cannot make antibodies to a structure which is seen as 'self'.

1.2 Cancer and heart disease

In developed countries cancer and heart disease are the major causes of death. They are both multifactorial diseases and therefore involve genetic predisposition and environmental factors. Although we can now identify genetic risk factors and environmental stimuli it is difficult to assess risk for an individual patient. Indeed, everyone seems to know someone who has smoked cigarettes all their life and lived into their 80s. Nevertheless, we all appreciate that smoking is a trigger in susceptible individuals and is the major cause of lung cancer. Presumably, if we could identify those with genetic predisposition we could more effectively target our anti-smoking campaigns.

1.2.1 Cancer

In the 1970s Richard Nixon, the president of the USA, stated that he would cure cancer in his term of office: huge sums of money were invested but no positive results were obtained. Indeed, cancer is probably as important today as a killer as it was 100 years ago. One problem faced by scientists involves understanding the mechanisms of cancer. However, although many cancers may now be defined with respect to mutations in oncogenes and tumour suppressor genes, most cancers involve numerous genetic mutations and determining which mutated protein, if any, will provide a suitable therapeutic target is difficult. Additionally, even though a mutation is detected an environmental stimulus is often a key factor without which the disease does not progress.

Currently the most effective treatment for cancer is surgery. This is only effective if the tumour has not metastasised and spread to other parts of the body. There are fears, however, that surgery itself may liberate cancerous cells into the host system, causing further problems. Chemotherapy and radiotherapy are useful but toxic and many of the agents used damage DNA in healthy cells, leading to fears that the treatment may well cause mutations which will cause different forms of cancer to develop later. Currently researchers are investigating new therapies that either stop cancers from growing or spreading or allow the immune system to recognise the cancers more effectively and destroy them. The following therapies are undergoing trials:

- turning off oncogenes

- adding tumour suppressor genes

- manipulating cell cycle

- killing cancer cell with cytotoxic drugs

- manipulating transcription

- using antibodies bound to toxins to kill cells

- using personal vaccines

- using recombinant cytokines

- delivering *tumour necrosis factor* genes.

1.2.2 Heart disease

Heart disease is responsible for 50% of 'natural' deaths in the USA. It is currently treated with drugs, diet, lifestyle management, surgery and transplantation. Heart disease is another multifactorial disease that requires both genetic predisposition factors and environmental triggers. Many of the triggers have been identified, and these include smoking, stress, high fat diet and salt intake. Campaigns addressing the reduction of risk factors in the population are largely ineffective. Current research is focused on the treatment of thromboses, with recombinant drugs such as tissue plasminogen activator factor, and the manipulation of animal organs for transplant to alleviate the current shortage of organs. Gene therapy has been attempted to treat familial hypercholesterolaemia.

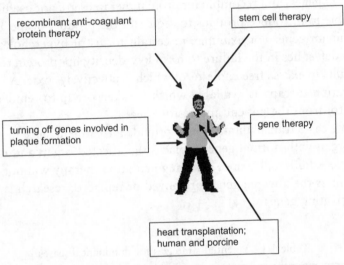

Figure 1.1   Therapies undergoing trials for heart disease

1.3 Genetic diseases

Although genetic diseases are relatively uncommon they are frequently incurable and often untreatable and therefore threaten the lives of sufferers. However, some diseases occur at high frequency within affected populations, for example thalassaemia in individuals of Mediterranean origin (1 in 20 carriers). Many problems associated with inherited disease may be overcome by using screening strategies: for example population screening and prenatal diagnosis where appropriate. However, abortion of affected fetuses is an ethical issue and many families do not realise they have an inherited disorder until they produce an affected child.

New therapies are currently being assessed and these range from drugs, recombinant proteins and gene therapy to immunotherapy. In designing treatments for inherited diseases we need to understand the basis of recessive and dominant diseases.

1.3.1 Dominant diseases

Dominant diseases require only one mutated gene for an affected individual to show symptoms and the bad gene always 'over-rides' the *good* gene. Examples of dominant disease include Huntington's disease and Marfan's syndrome. This type of disease can only be treated by removal of the bad gene or its product, therefore gene therapy, which currently involves the addition of genes, and recombinant protein therapies are not useful. It may be possible to introduce proteins to sequester the product of the action of the dominant gene. For example, in certain forms of hypercholesterolaemia, the defect lies in the failure to form low-density lipoprotein receptors. This results in excess free cholesterol which is effectively toxic. A chemical sequestration therapy is available which is taken orally to remove excess cholesterol from the patient and restore health.

In some cases transplantation may provide a treatment since in this case a new organ with normal genes replaces the affected organ and its defective genes. Additionally, since gene augmentation therapy will not provide a cure unless the affected gene is removed or replaced, research is focused on deactivating genes.

Table 1.1   Potential strategies for dominant diseases

- turn off genes by mutagenesis – not yet feasible in humans
- gene correction by homologous recombination; some success in mice
- germline therapy
- turn off genes using oligonucleotides; need constant supply
- remove poison gene product
- manipulate gene regulation

1.3.2 Recessive diseases

Recessive diseases, such as cystic fibrosis and sickle cell disease, are theoretically easier to treat since in the presence of one good gene the symptoms do not appear. These diseases are good candidates for gene therapy and transplantation since there is no requirement to remove the defective gene. Recombinant protein therapy may be useful in cases where delivery of the protein into the blood stream, intestinal tract or lungs will effect a cure. For example, Haemophilia A is caused by a fault in the production of the clotting factor, Factor VIII. This can be provided as a recombinant protein into the blood stream and effect a cure. However, the *CFTR* gene that is affected in cystic fibrosis produces a protein that must lodge within the cell membrane. Recombinant protein therapy could deliver the protein to the cell, but the protein would not be taken up into the membrane to perform a function and cure the disease.

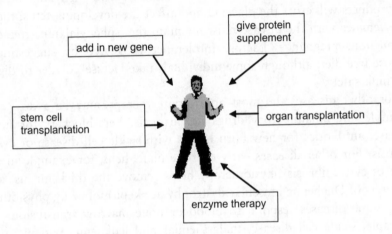

Figure 1.2 Strategies for recessive diseases

1.4 Role of molecular biology in therapeutics

Advances in molecular biology have facilitated diagnosis of disease at the molecular level: such an understanding of the disease process may suggest a treatment strategy. Recombinant DNA technology may be used to produce therapeutic proteins, transgenic animals, cloned cells or gene constructs for gene therapy. An understanding of the molecular biology and genetics of a disease are vital to allow us to decide which 'mutant' genes to target. In infectious disease, an understanding of microbial genetics and molecular biology allows us to formulate vaccines and molecular therapies. Once the mechanism of infection has been clarified, novel drugs can be developed.

However, for many diseases we do not need molecular therapeutics. In these cases we use conventional methods to treat the patient. We do not achieve a cure but we improve the life of the patient. If a disease causes a reaction to a particular foodstuff we would advise avoidance rather than gene therapy. In the case of phenylketonuria we can treat the patient by suggesting a phenylalanine-free diet, which prevents the damage caused by the accumulation of the breakdown products of phenylalanine. The diet in these cases must be maintained until puberty when the organs are fully developed. At this stage most affected individuals are able to eat normally again. However, we must be aware of potential problems to fetuses carried by affected females. The phenylalanine and by-products will cross the placenta and affect the developing fetus; therefore females with the disease must maintain their phenylalanine-free diet throughout pregnancy. Lactose intolerance is treated by suggesting a lactose-free diet, although some individuals take lactase powder to digest the milk safely.

For other diseases the most appropriate therapy may be a drug, for example anti-inflammatory drugs for arthritis, beta-blockers for heart disease, antibiotics for new born babies with sickle cell disease or cystic fibrosis. For other diseases early exercise plans help, for example in the case of cystic fibrosis physiotherapy helps remove the thick mucus, and sufferers of Duchenne muscular dystrophy are kept mobile by physiotherapy. Some diseases respond to blood or bone marrow transfusion, for example, sickle cell disease, thalassaemias and leukaemia patients after chemotherapy. Other patients are treated by organ transplant; for example, cystic fibrosis may be treated by heart–lung transplant.

Table 1.2 Uses of molecular biology in therapeutics

- diagnosis of the disease at the molecular level
- production of therapeutic proteins
- identification of the proteins involved in a disease process
- production of gene constructs for gene therapy
- understanding of a disease; inform which mutant genes to target
- formulation of vaccines and molecular therapies
- animal models may yield clues to therapeutic strategies
- cloning may allow us to rapidly increase animal models
- therapeutic cloning may allow production of stem cells for organ regeneration
- manipulation of rejection targets in transgenic animals to produce xenotransplants
- recombinant products may be used *ex vivo* to stimulate cell

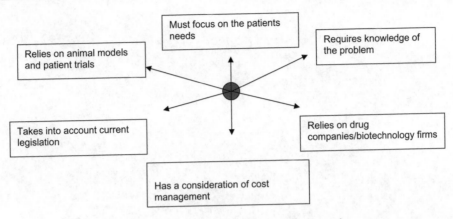

Figure 1.3 A rational approach to therapeutics

Points to consider

Ask your friends about the diseases covered. Are they well informed on
   the novel technologies?
Do you think such therapies will ever be used in developing countries?[5,6]
What problems do we face when treating genetic disease?

Notes

1 www.who.int/mdg/goals/goal6/en/
2 http://whqlibdoc.who.int/publications/2005/9241562986_part4.pdf
3 Coffin, J.M. et al. (1997) *Retroviruses*. Cold Spring Harbor Laboratory Press,
   New York
4 Ikuta, K. et al. Positive and negative aspects of the human immunodeficiency
   virus protease: development of inhibitors versus its role in AIDS pathogenesis.
   Microbiol. Mol. Biol. Rev., 2000, **64**, 725–45
5 Acharya, T. et al. Biotechnology to improve health in developing countries – a
   Review, *Mem. Inst. Oswaldo Cruz.*, 2004, **99**, 341–50
6 www.ccne-ethique.fr/english/avis/a_046p02.htm

Points to consider

Notes

2

Prenatal diagnosis and pre-implantation

2.1 Should we treat inherited diseases?

In theory 'prevention is better than cure' seems a sensible answer to the problem. However, screening for genetic diseases raises many issues that would need to be addressed before we could imagine that all diseases could be prevented.[7] For many genetic diseases there is not a strict relationship between the genotype (the base sequence of the gene), and the phenotype (the disease symptoms and severity). Two factors that play a role in many diseases are penetrance and expressivity.

Penetrance suggests that within a population with the disease gene only a proportion of those who inherit the gene will actually show symptoms. Since this effect is only seen in dominant diseases we know that it is not a function of the number of bad gene copies. Thus, although a fetus may be diagnosed as having the rogue gene it may never develop symptoms. Expressivity on the other hand relates to the severity of symptoms. In some diseases patients may be either mildly affected or so severely affected that their lives are foreshortened. Since the precise mechanism is not understood there are a range of diseases for which we cannot predict phenotype by analysing genotype. Additionally, a screening programme is reliant on an assumption of Mendelian transmission of disease. In some diseases, such as Duchenne muscular dystrophy (DMD), spontaneous mutation accounts for more than 1 in 3 of the cases recorded. Since most prenatal diagnosis is currently carried out on at-risk families with a genetic history of the disease, spontaneous mutations would be missed.

Molecular Therapeutics: 21st-century Medicine by Pamela Greenwell and Michelle McCulley.
© 2007 John Wiley & Sons, Ltd

2.2 Genetic screening

Theoretically to 'rid the world of genetic disease' we would have to screen everyone for every known disease. The numbers of affected individuals detected would be minuscule and the cost prohibitive. Having identified carriers of disease genes, we would then need to set up a prenatal screen of subsequent offspring. We cannot test every fetus as the procedure itself has a risk of triggering a miscarriage in 1–2% of cases. At this stage, any affected fetuses would be aborted – a difficult decision for most individuals. If we truly wished to eradicate inherited disease, we would also abort all carriers of recessive and X-linked diseases, even though these would be phenotypically normal. It is unlikely that anti-abortion and pro-life lobbies would agree to such a move and most of the public would feel uneasy with such an approach. We would also need to decide which diseases were to be detected. Should we test only for those which are immediately life-threatening to a newborn? If so, we would test for very few diseases and would save little money, as these children would die rapidly and not present a drain on our resources. How many inherited diseases are immediately life-threatening? The answer is very few; anencephaly is one example. Many of the diseases which cause still-birth are genetic but not inherited, for example complete or partial trisomies (with the exception of chromosome 21, X and Y), monosomies (with the exception of X). These would only be detected if every fetus were analysed for genetic defects in the womb.

Perhaps we should qualify our criteria and look at diseases which severely affect quality or quantity of life. We would then include diseases such as DMD. However, we would struggle with sickle cell disease (SCD), which can present as a mild condition and is treatable, or with cystic fibrosis (CF) or inherited cancer where patients can live into their 30s. Are we suggesting that the life of a child with CF is not worth saving and that we should save money by aborting affected fetuses rather than developing treatments for such diseases? Who would decide which disease to test for? How would this be implemented and what would be the consequences of refusal for testing? Within families who know they have a high risk of having children with genetic abnormalities, one would like to imagine that common sense would play a role in family planning; sadly evidence suggests that the drive to reproduce frequently outweighs the risk.[8,9]

Even if we could test everyone, adults, children and fetuses, and solve the problems of understanding the relationship between genotype and phenotype for every disease we would still face problems of compliance. Many religious groups abhor abortions and it would be difficult to legislate to force abortion, although this has been successful in the control of thalassaemia in Sardinia, a Catholic country. Until the introduction of screening programmes most of the Sardinian health budget was spent treating sufferers. A mass screening scheme was set up to identify carriers and a law was instituted making prenatal diagnosis mandatory for two carriers who wished to have children. Any couple who then had an affected child was made responsible for the upbringing of the child and its medical treatment. The effect was dramatic and the incidence of the disease has fallen to zero in Sardinia. Whether this is morally or ethically desirable is an interesting question. More recently a pre-implantation genetic diagnosis scheme (see below) has been instituted which should help prevent abortion.[10] Since we currently test only for life-threatening diseases within at-risk populations there appears to be little problem. However, in China there have been rumours of eugenics programmes being instituted, forbidding those with genetic disease and those with 'mental diseases' from having children.[11] Is this the beginning of a 'slippery slope'?

There is also the problem of late-onset disease, such as Huntington's disease (HD) and inherited forms of Alzheimer's disease (AD), where sufferers are unaware of the problem until they have already reproduced and passed on the genes. It is an ethical minefield to suggest prenatal diagnosis for such diseases when a newborn may have forty years of good life ahead. Who could be sure that a cure would not be found in that time?[12]

The cost of treating inherited diseases is disproportionate to the number of sufferers and issues such as the allocation of limited resources and the rights of patients to expensive therapy must be addressed. Nevertheless, it would be immoral to suggest that these diseases be ignored since many result in pain and suffering and untimely death.

2.2.1 Pre-implantation genetic diagnosis

A perceived solution would be pre-implantation genetic diagnosis (PIGD) in which cells from a fertilised egg produced by *in vitro* fertilisation (IVF) are tested for specific mutations and chromosomal disorders and only those free of disease are then implanted.[13] Until recently the number of mutations that could be investigated in a single sample was small (5 or 6) as the test was carried out on a single cell. With the advent of new technology such as whole genome amplification and comparative genome analysis, more tests can now be performed.[14] Nevertheless, this is an expensive technology requiring skilled personnel and results are normally backed up by prenatal diagnosis (PND). It is inconceivable that every family with inherited diseases could make use of such technology even in the UK where health-care provision is well supported. One of the obvious problems is that currently only those known to be at risk by virtue of family history or the birth of an affected child are tested. We would therefore need a mass screening programme as a first move to identify those at risk of transmitting inherited diseases and then persuade them to undertake PIGD to ensure 'normal' pregnancy. Using this strategy we would still miss the spontaneous mutations and hence we would still have children born with diseases such as DMD, Huntington's disease and achrondroplastic dwarfism.

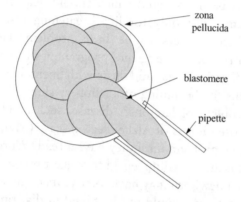

Pre-implantation genetic diagnosis requires cleavage-stage embryo biopsy and involves first making a hole in the zona pellucida and then gently aspirating the cells one at a time out of the embryo. These cells are then used for cytogenetic or polymovase chain reaction analysis.

Figure 2.1   Cleavage-stage embryo biopsy

PIGD has been linked in the press with 'designer babies', the idea that we could pick and choose qualities in our offspring.[15] This is naïve as we are not changing the genetic character of the babies but simply selecting from a very limited pool those free of a single mutated gene. There is a limit to the variability of offspring of one couple and such variability is random. Thus if 10 eggs were harvested, fertilised and tested they would on average each have half the genetic information of each parent and like siblings would share many characteristics.[16] Thus, it is impossible to imagine a world without inherited disease; we could reduce the burden by education and selective screening but we cannot prevent the birth of every affected child.

2.3 Counselling

It is difficult to predict the potential impact of therapies that are under-going trials. What can one tell parents of a fetus known to have CF? Do we tell them about gene therapy, xenotransplantation and stem cell work? If we do they may decide not to abort. If the therapies do not come to fruition their hopes and aspirations for the child will be dashed. If on the other hand we fail to inform parents, they may abort today, and within the next 12 months a therapy may be developed that would have saved their child's life. Can you imagine the guilt, knowing that you had aborted a child whose life could have been near-normal with the appropri-ate therapy? It is clearly a no-win situation and currently we try to give as much information as possible to allow parents to make the 'right' decision.

There are guidelines set down for genetic testing and for counselling of families with inherited disease and parents of affected children.[17,18] It is worth remembering that for many families, they are only aware that they have a problem following the birth of their first affected child. Thus, under the current testing regimen, we will always have affected children for whom treatment will need to be developed. However, if we give normality to affected individuals, but we do not change their germline DNA, there is an argument that all we are doing is encouraging bad genes to be passed on. It is clear from the experience of treating haemophiliacs that once a patient is treated and regains 'normality', they want to enjoy life and part of that is to have children. We are therefore, albeit in a small way, tampering with evolution and ensuring 'survival of the least fit'.

Points to consider

Talk to you friends and families about the issues.
Can you really expect people to think about genetic disease before having children?
In terms of education, using your own experience, do we inform people about these problems? If not, how would you do this?
Check out the campaigns for prevention of SCD: these are targeted and work!

Notes

7 www.nature.com/ejhg/journal/v11/n2s/pdf/5201113a.pdf

8 http://content.nejm.org/cgi/content/full/347/23/1867

9 www.nlm.nih.gov/medlineplus/genetictesting.html

10 Monni, G. et al. Preimplantation genetic diagnosis for beta-thalassaemia: the Sardinian experience. *Prenat. Diagn.*, 2004, **24**, 949–54

11 Su, B. and Macer, D.R. A sense of autonomy is preserved under Chinese reproductive policies. *New Genet Soc.*, 2005, **24**, 15–29

12 Cox, S. et al. Prenatal diagnosis for late-onset disease – always room for discussion. *Fertility and Sterility*, 2005, **83**, 1887–8

13 Draper, H. and Chadwick, R. Beware! Preimplantation genetic diagnosis may solve some old problems but it also raises new ones. *Med. Ethics.*, 1999, **25**, 114–20

14 Handyside, A.H. et al. Isothermal whole genome amplification from single and small numbers of cells: a new era for preimplantation genetic diagnosis of inherited disease. *Mol. Hum. Reprod.*, 2004, **10**, 767–72

15 Edwards, R.G. Ethics of PGD: thoughts on the consequences of typing HLA embryos. *Reprod. Biomed.*, 2004, **9**, 222–4

16 http://humrep.oxfordjournals.org/cgi/reprint/18/7/1368

17 Knoppers, B.M. and Isasi, R.M. Regulatory approaches to reproductive genetic testing. *Hum. Reprod.*, 2004, **19**, 2695–701

18 www.pubmedcentral.gov/picrender.fcgi?tool=pmcentrez&blobtype=pdf&artid=228480

3

Simple protein replacement therapy

Replacement protein therapy is one of the oldest treatments for genetic disease. The rationale is that if a patient lacks a protein that can be supplied through the blood, intramuscularly, as an aerosol or into the digestive tract, we can give the patient protein therapy. Until recently most therapeutic proteins were derived from animal or human tissue. For example, Factor VIII, the clotting factor missing in haemophilia A sufferers, was obtained from human plasma, insulin was isolated from pig pancreas and growth hormones were purified from human cadaver tissues.

However, there have been numerous problems in the last two decades. Initially, the demand for Factor VIII was greater than the supply from the blood products laboratories in the UK and therefore extra factor had to be obtained from the USA. In the USA there was a paid donor policy which tended to encourage donations from the poor, down-and-outs and drug addicts.[19] Many of these were infected with HIV. Prior to the identification of HIV as a causative agent in AIDS, many haemophiliacs received contaminated blood products. Subsequent treatment of pooled plasma to remove the virus and a careful donor screening campaign have reduced the risk but other viruses have been identified as possible contaminants of blood-derived proteins, for example hepatitis C. Additionally, in the UK, 6% of those who received growth hormone or fertility hormones derived from human sources were infected with Creutzfeld–Jacob disease (human spongiform encephalopathy) and have since died.

Current fears about the transmission of infections from animals to humans cast doubts on the safety of all animal-derived material. Indeed,

Molecular Therapeutics: 21st-century Medicine by Pamela Greenwell and Michelle McCulley.
© 2007 John Wiley & Sons, Ltd

a recent incident involving pooled plasma from the USA has highlighted a major problem. Pools of 512 donations are routinely tested for hepatitis (A, B and C) and HIV, then thousands of negative donations are pooled for extraction of proteins. In the USA, an independent study of 1024 donors whose plasma made up 2 separate pools sent to and used by the UK showed that 2 samples, one in each pool, tested positive by polymerase chain reaction (PCR) for hepatitis B. What can we do to ensure safety?

3.1 Preventing transfusion-transmissible infectious diseases in the UK[20]

In the UK, blood donation is unpaid and voluntary. The National Blood Service is responsible for the selection of donors. This includes ensuring that the donor is provided with clear, understandable and up-to-date information and also ensuring that the donor has understood this information. Every donor must complete a questionnaire about their health and is interviewed by a health professional to assess their exposure to any transfusion-transmissible infection. The following are judged to be risk factors: major surgery; tattoo; body piercing; acupuncture; close household contact with persons with certain infectious diseases. The world is now easily accessible by air travel and this may lead to development of 'exotic diseases' with asymptomatic people donating infectious blood. A detailed travel history is now obtained from all donors to minimise the risk of transmission of, for example, malaria and West Nile virus. The following infectious diseases result in permanent deferral of donation: hepatitis B; hepatitis C; HIV 1 and 2; HTLV I/II; babesiosis; kala-azar (visceral leishmaniasis); chikungunya virus; methicillin-resistant *Staphylococcus aureus*; and *Trypanosomiasis cruzi* (Chagas disease).

In addition, those identified as having an increased risk of developing a prion-associated disease are permanently excluded from donation. This includes those who have received human pituitary-derived hormones, human dura mate or corneal or scleral grafts; those at risk of inherited prion diseases; recipients of blood or blood products since 1980. Given the incidence, albeit low, of new-variant CJD in the UK, the National Blood Service recognised that there was a risk of CJD transmission via blood and blood products. In the UK blood for transfusion is depleted of white cells to reduce risk and blood for the isolation of proteins is sourced from outside the UK. Prion disease (new-variant CJD) is known to have been passed on to women who were given fertility hormones and children given human growth hormone.[21,22] Interestingly, many of the hormones were derived from cadavers from countries in which new-variant CJD has never been reported and where the incidence of inherited or sporadic prion diseases is less than 1 in a million. The exact mechanism whereby contamination with prions occurred is unknown, although cadaveric materials were pooled and it is known that pooled material can present greater risks than material derived from a single cadaver. Abnormal prions, even at low level appear to act to change the nature of the normal prion proteins present.

Parenteral drug users with a history of non-prescribed drug use, including body-building steroids or hormones, are banned from donation, as are those whose sexual behaviour puts them at high risk of acquiring severe infectious diseases that can be transmitted by blood. The latter includes prostitutes, men who have sex with other men, and those who have had sexual intercourse with inhabitants of areas where HIV is endemic. Xenotransplant recipients and their sexual partners, are also permanently deferred. In addition those who have suffered malignant neoplasms are also banned from donation.[23] WHO is now worried about transmission of leishmaniasis by blood transfusion. The disease is normally spread by sandflies so epidemiological studies are difficult. However, a recent survey showed that 61 of 463 blood donors in Monaco who were asymptomatic tested positive for the disease.

3.2 Ensuring the safety of organ transplants

In theory, assuring safety of blood products is relatively easy, as the donor can be interrogated. Although it is difficult to ensure that the information given is correct. However, this is much more of a problem when material is obtained from cadavers. In this situation we are reliant on relatives for information at a time when they are distressed. Often close relatives are unaware of 'unsafe practices' that would preclude their relative from tissue donation. Some organs do not remain viable for an extended time and it is unthinkable to insist that an organ donor be kept alive until all possible health checks have been undertaken. There are problems associated with supply of organs and tissues and compliance of relatives for the use of that tissue. The issues of infection are the same as those seen with transfused blood.[24] Additionally we need to be aware of the potential of transferring malignant cells if whole tissue is used.

3.3 Preventing transfusion-transmissible infectious diseases worldwide[25,26,27,28]

The World Health Organization (WHO) regularly collects global data about the safety and availability of blood for transfusion. The objective of these surveys is to measure progress in the 192 WHO member states and 2 associate members towards a sustainable and safe blood supply. Countries provide data on:

- Number of blood units collected
- Unpaid voluntary blood donations
- Family replacement donors
- Paid donors
- Blood screened for major infections.

Data for 2001–2 were received from 178 of the 192 member states of WHO (representing 95% of the world's population). The summary report and a copy of the questionnaire are available.[29]

3.4 HIV

HIV is a particular problem associated with blood products. It has been estimated that the risk of contracting HIV from a contaminated product is 96%.[30] This should be compared with the 1% risk of acquiring HIV through heterosexual sex with a single infected partner. Indeed, in the UK there are no reports of transmission of HIV from infected haemophiliacs to their spouses. Since the recognition of the problem, the transmission of HIV via blood has reduced dramatically and in the USA it has been stated that there are now only 2 cases of HIV for every million units banked. This makes the risk small to those receiving one-off transfusions but haemophiliacs received material from pooled blood and therefore products derived from thousands of donors, thus increasing their risk. Figures have also been produced for the risk of other viral infections from blood transfusion. The overall risk of contracting HIV, hepatitis B and C and lymphotrophic virus are 1 in 34,000 units of blood.

However, in some developing countries there is a much greater problem. In Democratic Republic of Congo, for example, 25% of individuals are thought to be HIV-positive. In areas of high viral load and poor testing facilities or shortage of donors, infected material is collected and used. Worldwide, transfusion is thought to account for 15–20% of HIV transmission. The problem of viral contamination is often ignored when material is isolated from human tissue. For example, in the early 1990s, a French company isolated albumin from human placentas, collected from 44 countries worldwide without first checking for HIV risk.

There are, however, advantages in using material isolated from human tissue. The product is less likely to be recognised as foreign since it has been derived from the same species although it is still non-self and may still be recognised as foreign by virtue of polymorphic changes in the protein. However, in the case of haemophilia A, some individuals with large gene deletions never produce Factor VIII and recognise the therapeutic material as non-self and raise antibodies to it. The product is also much cheaper than the recombinant equivalent since methods of isolation have already been developed and there is little capital outlay to recoup. Nevertheless, the problems associated with disease transmission via blood- and tissue-derived products have resulted in the development of recombinant alternatives to these natural proteins.

Points to consider

How can we ensure human-derived material is safe?

Notes

19 Starr, D. (1998) *Blood: An Epic History of Medicine and Commerce*. Alfred A. Knopf, New York
20 www.hpa.org.uk/CDR/archives/archive04/news/news1204.htm
21 http://jnnp.bmjjournals.com/cgi/reprint/72/6/792
22 Lindholm, J. Growth hormone: historical notes. *Pituitary*, 2006, **9**, 5–10
23 www.transfusionguidelines.org.uk
24 www.eurosurveillance.org/eq/2005/01-05/pdf/eq_2_2005_17-19.pdf
25 Ahmed, S.G. Laboratory strategic defense initiatives against transmission of human immune deficiency virus in blood and blood products. *Niger. Postgrad. Med. J.*, 2003, **10**, 254–9
26 www.blackwell-synergy.com/doi/abs/10.1046/j.1365-3148.2003.00453.x

27 www.blackwell-synergy.com/doi/abs/10.1046/j.1365-3156.2000.00621.x
28 Moore, A. *et al.* Estimated risk of HIV transmission by blood transfusion in Kenya. *Lancet*, 2001, **358**, 657–60
29 www.who.int/bloodsafety/global_database/en/
30 Baggaley, R.F. et al. Risk of HIV-1 transmission for parenteral exposure and blood transfusion: a systematic review and meta-analysis. *AIDS*, 2006, **20**(6), 805–12

4

Recombinant protein production

Recombinant proteins are proteins synthesised by expression of a cloned gene in another species or in cell culture. The production of the protein can be achieved in a number of different organisms: these include bacteria, yeast, insect and mammalian cells. The choice of host organism generally depends on the complexity of the product to be produced.

In order to make a recombinant protein, the gene encoding the protein must have been cloned and characterised. The gene must then be sub-cloned into an appropriate expression vector and used to transfect cells. The nature of the vector is dependent on the host in which the gene is to be expressed. To be therapeutically useful, the product must retain its biological activity and tests should be available to assess the function and biocompatability of the product. Simply testing a protein *in vitro* will not highlight problems associated with immune recognition of the product by the patient. Similarly, a product that has been designed as a therapeutic agent for use in humans will not always fare well when tested in animal models.

4.1 Choice of organism

The organism that will express the recombinant protein must be easy to transfect and lack viruses or toxins that may affect the patient. Since cost of the product will be a consideration, the process must be easy to scale

Molecular Therapeutics: 21st-century Medicine by Pamela Greenwell and Michelle McCulley.
© 2007 John Wiley & Sons, Ltd

up and the product should be purified readily. Fermentors of greater than 1000 litres capacity are used to grow organisms on an industrial scale. The choice of organism is influenced by the characteristics of the vector that can be used. The size of the protein product is therefore a consideration since different vectors have capacity for different size inserts. The complexity of protein processing and the potential toxicity of the product to the host organism are also considerations.

The vector should be able to stably incorporate the foreign DNA. The incorporation of a selectable marker ensures that all the cells that are grown contain the foreign gene. It is usual to incorporate a gene conferring resistance to an antibiotic in bacterial vectors and primary selection will be based on growth in selective media, the assumption being that antibiotic-sensitive cells taking up the vector will gain both the functional gene and the antibiotic resistance. From an environmental and safety aspect, the vector should not be able to escape from the cultured cells and pose an environmental risk.

Table 4.1   What is needed to make a recombinant protein?

The tools:
- the gene encoding the protein of interest +/− promoters
- a vector, a carrier of the gene, specific to the organism in use and big enough to carry the gene
- the organism to be transfected with the vector
- growth media, fermentors and downstream processing (purification) facilities

The organism:
- must be easy to transfect
- must produce active protein
- must not produce protein which is immunogenic to the recipient
- the product must be free of viruses and toxins
- the process must be flexible and easy to scale up
- downstream processing should be easy
- cost is a consideration in industrial-scale production

What influences our choice of organism?
- the size of the protein product
- the complexity of the protein processing
- the potential toxicity of the product to the host organism
- cost

Why is size of the gene a problem?
- determines the type of vector that can be used
- in *E. coli*, plasmids can harbour up to 6 kb (kilobases) of foreign DNA, cosmids up to 50 kb
- in yeast we can use plasmids or YACs (yeast artificial chromosomes)

Table 4.1 Continued

- in mammalian cells we can use SV40 virus, YACs and HACs (human artificial chromosomes)
- in insect cells we use baculoviruses

What features of the vector are important?
- it should be able to incorporate stably the foreign DNA
- it should have a selectable marker to ensure that all cells which grown contain the foreign gene
- it should be easy to clone the gene into the vector in the 'right' orientation and reading frame
- the vector should allow production of large amounts of stable product
- the vector be unable to escape from the cultured cells i.e. should not pose an environmental risk

The simplest system for production of recombinants uses *Escherichia coli*. *E. coli* has a number of advantages: it is a well-studied organism and many vectors are available. The bacterium grows quickly in culture, produces large amounts of products, is economical to grow and multiplies rapidly. However, the gene must be inserted in the form of cDNA since the bacterium cannot splice out introns and signalling requirements must be included in the construct to ensure that expression occurs appropriately. Additionally, *E. coli* does not process proteins in the same way as humans and therefore is unable to add functional groups such as sugars to the protein products. Since many proteins are reliant on such processing for their activity this is obviously a problem. Additionally, the foreign sequence may cause premature termination of synthesis in the bacterial host and the protein once made may be degraded by the bacterium. Codon usage may also pose problems since different organisms have different preferences in the use of some codons.[31] Each organism possesses the appropriate types of tRNA molecules for its natural codon usage and may not be able to cope with 'novel' uses of the triplet code.

In *E. coli* there is a choice of promoters that can be used to facilitate expression of the foreign gene. Strong promoters give continuous high-level expression of the protein product. This may be useful in that large amounts of the product may accumulate in the growth medium for collection and purification. However, if the product is toxic to the host cells, then a promoter can be used to regulate the expression of the protein. For example, the foreign gene could be cloned into the *lac* operon. No products are formed from this promoter under normal growth conditions. However, when lactose is supplied as a sole carbon source the promoter is activated and gene expression is turned on and the protein produced. Such an inducible system will allow production of product at a set time in the fermentation.

If we clone into the *lac* gene we may synthesise a fusion protein that contains both β-galactosidase from the host and the foreign protein. This is advantageous in terms of product stability in that fusion proteins are less likely to be destroyed by the *E. coli* proteases than by the foreign protein alone. Additionally, the bacterial protein acts as a carrier for the foreign protein, allowing secretion of the protein products into the medium. The *E. coli* β-galactosidase is a large protein of over 110 kDa and it is one of the largest proteins produced by the organism. Fusion proteins will be far larger than the native protein – this could facilitate purification. Additionally, there are commercially available affinity gels for the purification of β-galactosidase that will also bind the fusion protein. Once eluted, the purified fusion protein can be chemically treated to remove the β-galactosidase sequence, thus freeing the foreign protein into solution. The mixture is then applied to the affinity ligand a second time. In this case the β-galactosidase will bind, but the foreign protein will not, eluting with the unbound material. Hence purification is simple.

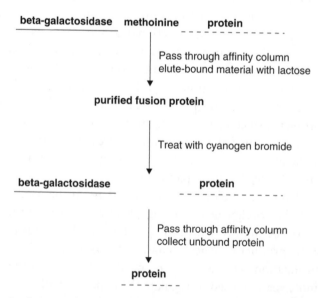

Figure 4.1 A schematic showing the purification of the recombinant protein using an affinity ligand for β-galactosidase

However, the production of a large fusion protein in which only a small fraction of the synthesised product is the recombinant protein is not cost-effective. As we will see in a subsequent example, the use of the *Trp* promoter results in a smaller fusion protein which is energetically more favourable to produce, resulting in a lower-cost product.

4.1.1 Somatostatin: an example of protein produced in *E. coli*

Somatostatin was the first recombinant protein to be synthesised in *E. coli* as a recombinant protein. The anti-growth hormone is small, having only 14 amino acids, and is active as a single-chain unprocessed polypeptide. Hence it was relatively simple to synthesise in *E. coli*. The strategy employed in its manufacture involved production of a β-galactosidase fusion protein, purification on an affinity ligand, chemical removal of the β-galactosidase moiety and affinity ligand chromatography. In this case the fusion protein was constructed so that the β-galactosidase and somatostatin polypeptides were joined together by a methionine residue. Treatment of the fusion product with cyanogen bromide cleaved the protein at the methionine residue, resulting in the production of free β-galactosidase and free somatostatin. The β-galactosidase was then removed by use of an affinity ligand. In terms of a marketable product, somatostatin was a poor choice. However, it was the model system that demonstrated that recombinant proteins could be synthesised in this type of system, and this preliminary study led to the manufacture of more relevant therapeutic proteins.

4.1.2 Insulin: an example of a recombinant protein[32]

The market for recombinant insulin is vast. Until the production of the recombinant protein, patients were given insulin purified from pig pancreatic tissue. The porcine insulin differs by one amino acid from the human, the last threonine being replaced by alanine, and caused no immune response. However, the use of animal material for the isolation of proteins raises a range of questions concerning safety, quality assurance and quality control. Therefore the interest of the biotechnology companies in the production of insulin was twofold: they would make a safer product which then would be commercially attractive.

Insulin is a simple protein, although more complex than somatostatin. *In vivo* it is synthesised as a proinsulin with three domains, A, B and C; the protein folds and forms disulphide bridges between the A and B domains and the C domain is cleaved out by proteases. The most common approach to the synthesis of insulin involves making the proinsulin as a recombinant, harvesting the product, and then treating with protease *in vitro*. In this case disulphide bridges form naturally in *E. coli* and the C domain is removed using an immobilised enzyme. Alternatively the A and B domains could be synthesised separately in two experiments and then chemically treated to allow formation of the disulphide bridges. Both approaches are used although the chemical treatment is seen to provide more hazards than the proteolytic approach.

The product is absolutely identical to human insulin. The *insulin* gene is cloned into the *TrpE* gene of the vector since as the tryptophan synthetase encoded is small, the amount of insulin made as a proportion of the total protein is higher than when the *lac* operon is used. If we clone into the *lac* operon we have one molecule of insulin produced for every 2059 amino acids synthesised whereas cloning into the *TrpE* gene produces one molecule of insulin for every 433 amino acids produced. Interestingly, athletes are now using recombinant insulin that is indistinguishable from normal endogenous insulin to increase glycogen stores in muscle before events. The insulin is quickly removed from the blood stream and it is impossible to determine whether an athlete has used this to boost performance.

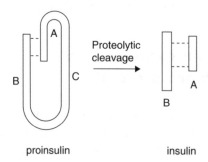

Figure 4.2 *In vivo* processing of insulin

'If the protein is active as a single unprocessed polypeptide then make it in *E. coli*.' This is the take-home message for the recombinant protein industry. *E. coli* systems allow production of large amounts of proteins cheaply. The process is simple to validate and well understood. Different variations of the process may be used to produce specific proteins. However, some proteins must be fully processed for activity. In general these cannot be made in *E. coli*. The processing of proteins may involve glycosylation: the addition of sugars to the protein chains in either N- or O-linkage, acetylation, phosphorylation, carboxylation and methylation of amino acids, proteolytic cleavage of the product or the formation of disulphide bonds.

4.2 Alternatives to *E. coli* for the production of recombinant proteins

There are a number of alternative systems for the production of recombinant proteins. These include yeast[33] and fungi,[34] insect cells and mammalian cells. Each system is useful for the production of some but not all proteins. Yeast was the first eukaryotic system investigated by researchers intent on production of recombinant proteins. Yeast had a number of advantages not least of which was the fact that yeast fermentation was an ancient skill used commonly by the brewing industry. Yeasts are relatively easy to grow and do not require complex and costly media. They are easy to transfect and there are a number of available vectors. Additionally, early workers felt that as the yeasts are eukaryotic and glycosylate their own proteins they would be able to modify human proteins producing active non-immunogenic products. However, it quickly became clear that although glycosylation does occur, the sugars added are not identical to those found in humans and the glycosylated proteins proved immunogenic and therefore of little practical use.

4.2.1 Insect cells[35]

Insect cells were also investigated as a host for the production of recombinant proteins. in theory, these cells have distinct advantages over other systems. Cells from *Spodoptera frugiperda* (the fall armyworm) grow readily in culture and are easy to transfect using insect-specific baculoviruses. Originally it was thought that these viruses provided no risk to humans; however, current work suggests that in culture, baculoviruses will infect human cells. The media for culture are expensive but serum-free, making downstream processing easier. The production of the product is inducible and the product is contained in inclusion bodies in the nucleus of the cell. The glycosylation patterns in insect cells differ dramatically from those seen in humans and thus proteins made in this system may cause immune reaction when used therapeutically.

4.2.2 Whole insects[36]

It has been suggested that whole insects could be used to produce recombinant proteins. A researcher described the production of 12 g of recombinant protein by transfecting insect grubs with virus and then allowing them to grow in a 'suitcase-sized incubator'. Each insect was provided with a stick containing a globe of nutrients on which it fed. At the end of the experiment the grubs were collected and minced and the protein was harvested. Although a novel experiment, it does not overcome the processing problems associated with insect cells.

4.2.3 Mammalian cells[37]

The problems described associated with the production of complex processed proteins have driven scientists to produce recombinant proteins in mammalian cells in culture. Mammalian cells are slow-growing and can be expensive in terms of the media required. The doubling times are often 16–24 hours and the risk of contamination by rapidly growing bacterial, fungal or yeast cells is high. However, they are easy to transfect and vectors are available to accommodate the largest of the human genes. The proteins produced are processed, but although glycosylation occurs it is not always appropriate (see 4.3.1) and hence the product may lack activity. Additionally, many cell lines harbour retroviruses that could be hazardous in a therapeutic product, for example the commonly grown HeLa (vaginal carcinoma) cell line harbours human papilloma virus (HPV).

4.2.4 Plants[38]

There has been an interest in producing recombinant proteins in plants in both developed and developing countries. We shall talk later about the production of vaccines in bananas, for example. In terms of production costs, plants are attractive options: they are cheap to grow and produce large amounts of product in a relatively labour-saving manner. However, they are not suitable for the production of complex processed proteins. In the wake of the GM (genetically modified) foods debate, many companies have abandoned ideas of making vaccines and recombinants in plants.

4.2.5 Transgenic animals

Transgenic animals, as we shall discuss in later chapters, are the products of germline gene therapy. They are manipulated at the egg stage to incorporate a gene of choice and then express the protein product in an easy-to-handle source such as milk. In the main, the technology is limited to sheep, goats, pigs and cattle. The major problems are associated with cost of production and development, the low levels of success achieved and the spectre of viruses and prions in the final product. Nevertheless recombinant proteins such as albumin, Factor IX and α-1-anti-trypsin are produced in this way. However, none have reached the market. A number of articles, for review see Powell,[39] have highlighted the fact that no transgenic product had been tested on humans and that there is little evidence that these products would actually work.

4.3 Problems with recombinant protein production

The major problems with the production of recombinant protein products relate to the delivery of DNA to a cell which then needs to adequately glycosylate and process the protein. DNA delivery or transfection may be achieved in a number of ways dependent on the organism. For example, calcium phosphate precipitation and electroporation, though inefficient, are adequate for transfection of E. coli. If fungal or plant cells are to be used there is an added step required to remove the thick cell wall and produce protoplasts prior to transfection. Insect cells are most usually transfected with baculovirus vectors. For mammalian cells, calcium phosphate and electroporation can be used but are inefficient. Direct injection of DNA is used in the production of transgenic animals where small numbers of eggs are injected – this is not practical when a large number of cells need to take up the gene. Viruses, such as the SV40 virus, have been used successfully to deliver DNA to cells in culture. There are safety implications associated with the use of viruses in mammalian cell culture and steps must be taken to ensure that the final product is virus-free.

4.3.1 Problems with glycosylation

Production of glycoproteins in animal cells is fraught with problems concerning correct glycosylation. In animal cell culture the polypeptide structure of the protein produced will be correct. We will not need to worry about codon usage, splicing of intron or processing. However, glycosylation still provides us with a major problem. Man and the higher apes are unique amongst mammals in that they do not produce the carbohydrate structure Galα1-3Gal as the terminal sugar residues on the ends of their glycoprotein chains. In fact the gene encoding the relevant enzyme required to synthesise this sugar is present but mutated in man and the higher apes, leading to its inactivation. Since this structure is not 'self', it is highly immunogenic to man and the higher apes, who produce natural antibodies to it. Thus, any recombinant protein produced in other animal species may contain this structure and, if present, the recombinant protein will be rapidly cleared from the system.

It is now possible to engineer animal cell lines to either inactivate the problem enzyme or to add in another enzyme that competes for the same substrate and produces the human blood group H antigen which itself is not immunogenic.[40] Current research suggests that, on addition of the human blood group *H* gene which encodes an α-2-fucosyltransferase (α2FT), the Galα1-3Gal epitope is suppressed.

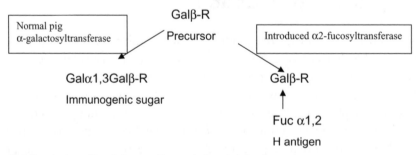

Gal: D-galactose; Fuc: L-fucose; R: remainder of molecule

Figure   4.3 Elimination   of   immunogenic   sugars   using   competing   α-2-fucosyltransferase

However, not only do species differ but also different cells within a species differ with respect to glycosylation! Thus, in a single individual, the glycosylation in kidney differs from that found in heart, which differs from that found in the gut.

Table 4.2   Tissue-specific expression of blood group antigens

Tissue	ABH antigen
Stomach (Se)*	+++++
Colon	+/−
Heart	−
Red cells	+++++

*Se: only 80% of Caucasians can produce H, and therefore A and B antigens in secretions

If we need to make Factor VIII which is normally produced in the liver, it might seem logical to use liver cells in culture to produce the product. However, not only is carbohydrate antigen expression tissue-specific, it is oncodevelopmental; that is, glycosylation changes from the embryo to the adult and in cancer and in cultured cells there is reversion to the embryonic pattern. Thus, liver cells in culture produce a glycosylation pattern identical to that of embryonic liver but different from that found in the adult liver. If we look at the example of carbohydrate blood group antigens, in cancer and in cell culture those cells that have normally expressed H, A and B lose antigens whereas those that did not originally express them gain antigens.

4.3.2 Effect of glycosylation

The roles of glycosylation are diverse. Some glycoproteins have sugar groups added to areas that are vulnerable to attack by proteases. The sugar in this case acts as a steric inhibitor. Carbohydrates carried on proteins of the cell membrane may be involved in bacterial and viral cell recognition. The sugar moieties also play a role in ensuring that the protein is correctly targeted, for example into the cell membrane or secreted from the cell. The sheer bulk of the added sugars may affect the conformation of the protein or the sugar itself may change the charge properties of the protein, facilitating changes in shape and activity. The presence or absence of sugars can also affect the functional activity of proteins in vivo. If we examine the case of Factor VIII: this protein is about twice as active in group A as in group O individuals. The difference in structure between these two blood group determinants is a single sugar: the A antigen has an additional α-N-acetylgalactosamine residue compared to the blood group O antigen.[41] If such a profound difference in activity can be achieved by the change in a single sugar, we can imagine the catastrophic effects ineffective glycosylation may have on some proteins. In industry, the production of the most active molecule possible will reduce wastage of capital, material and time.[42]

Gal: D-galactose; Fuc: L-fucose; GalNac: N-acetyl-D-galactosamine; R: remainder of molecule

Figure 4.4   The difference between blood group A and H antigens

A series of experiments carried out by Oxford Glycosystems showed the effect of different culture methods on the glycosylation and activity of IgM antibodies. They produced a single IgM species in mice and harvested serum, injected hybridoma cells derived from the mouse into the peritoneal cavity of a mouse to produce an ascites tumour which was drained to harvest antibody, and in an airlift fermentor (airlift 1). An airlift fermentor uses air to oxygenate and stir the culture contents and is used extensively in industry. The products, serum, ascites and airlift 1 had exactly the same amino acid composition but different glycosylation. A different IgM antibody was also produced in the airlift fermentor under identical conditions (airlift 2): this should have the same glycosylation as airlift 1 but a different amino acid structure. It is clear from the table that the airlift 1 product differed in glycosylation from the serum and ascites product despite the fact that the protein structure was identical.

Table 4.3   Glycosylation and IgM production

Sugar	Serum IgM	Airlift 1	Airlift 2	Ascites
	Sugar analysis of IgM %sugars			
Fucose	4	8	7	5
Galactose	12	14	16	14
Mannose	43	28	27	40
GlcNAc	33	21	32	34
Sialic acid	8	9	8	7
Other sugars		20	10	

The airlift 2 product also differed from airlift 1 in the amount of GlcNAc added to the recombinant. The half-life of the products also varied. The IgMs from serum, ascites and airlift 1 had half-lives of 14 hours, 13.2

hours and 1.3 hours respectively. This shows the profound effect of glycosylation on the half-life of the product. Interestingly, the half-life for the product from airlift 2 was 12 hours. The serum, ascites and airlift 1 differ with respect to half-life due to sugar composition but airlift 1 and 2 are different IgMs. This work suggests that every IgM species must be assessed separately and that no one culture method is a 'gold standard'.

Culture pH also affects expression rates and glycosylation of recombinant proteins. This presumably reflects the activities of the glycosyltransferases, which synthesise the sugar chains, each of which may have a different pH optimum. Studies comparing N-glycosylation of interferon under different conditions show profound effects of host cell type on N-glycosylation. Studies on tissue plasminogen activator factor (TPA) have highlighted the fact that a product that appears fully functional and active *in vitro* may not prove to be the best product for use *in vivo*. In this case, trials suggested one product had three times greater activity when measured *in vitro* whereas *in vivo* it was only one-seventh as active as its rival product. This presumably reflects clearance rates.

The results of testing products destined for use in humans must be interpreted carefully. A recombinant protein that produces no immune response in a mouse may, by virtue of the Galα1-3Gal epitope, be cleared readily from human systems, whereas a product that causes immune reaction in mouse may be completely harmless in humans.

4.3.3 Erythropoietin: an example of protein produced in mammalian cells

Erythropoietin is an interesting example of a recombinant protein, since prior to production as a recombinant the protein could only be obtained in milligram quantities from human urine, and thus its therapeutic potential was limited. It is also interesting since the product does not aim to cure any disease, but only alleviate symptoms. Nevertheless, it is said to have been the greatest breakthrough in clinical nephrology in the last 25 years. The protein stimulates the production of red cells and therefore may be used to treat anaemia in patents undergoing kidney dialysis. The product is said to improve the quality of life, increase appetite and vitality, and instil a sense of well-being in the patient. It is also used illegally by athletes to boost performance. However, subtle changes in the recombinant product have been used to detect those using the drug illegally.[43]

4.3.4 Production method[44]

The protein was purified from human urine and limited N-terminal sequencing was carried out. The gene was cloned by screening libraries with a degenerate probe. The cloned gene was then manipulated to produce a construct with the gene and a selectable marker which were then cloned into the SV40 viral vector which was then allowed to infect Chinese hamster ovary (CHO) cells. The selectable marker is an important addition. Cells taking up the vector need to be isolated from those wild-type cells in which no gene has been incorporated. If a selectable system is not used, the wild-type cells may have a growth advantage and protein production from the manipulated cells may be low. The selectable marker is often an enzyme that is taken up by the host cell lacking that enzyme in the same vector molecule as the therapeutic gene. Cells taking up the vector will then grow on selective media where the presence of the marker enzyme is required for survival. If the cell takes up the vector and marker it must also contain the therapeutic gene.

Once the cells had been selected they were grown in large-scale culture in fermentors and the product was harvested, characterised and purified. The product was then analysed for the presence of viruses and hamster proteins. Once made, the product underwent trials before marketing. Since the product is 165 amino acids in length, highly glycosylated, animal cells were the only feasible vehicle for production. CHO cells were selected as there are a variety of glycosylation mutants available for the production of adequately glycosylated proteins.

Table 4.4   The process for eukaryotic expression

1. Clone gene manipulated to add strong promoter and selectable marker, sub-cloned into SV40 vector
2. Introduce into Chinese hamster ovary cells (CHO) with a selectable marker
3. Grow on selection media to select expressing cells
4. Grow cells in large-scale culture
5. Harvest and purify product
6. Check for the presence of hamster proteins and viruses

4.3.5 Preparation of Factor VIII

The next obvious target for the biotechnology industry was the production of Factor VIII[45] as a recombinant.[46] It has been estimated that the sales have exceeded $4.6 bn. The problems associated with HIV and hepatitis C transmission from blood-derived Factor VIII and the current scares over the transmission of CJD have resulted in a drive to produce Factor VIII as a recombinant and to then prescribe the recombinant in preference to the blood-derived material. However, there are severe problems in the production of Factor VIII as a recombinant protein. The gene encoding Factor VIII is amongst the largest of the human genes. The genomic DNA is 186 kb with 26 introns, the mRNA is 9 kb. The protein has 17 disulphide bridges and is heavily glycosylated. The glycosylation is known to affect the activity of the molecule. This was obviously a difficult problem for the biotechnology industry and the product was only developed after much research and expense. The preliminary production was similar to that used for erythropoietin. CHO cells were infected with a viral vector containing the gene for Factor VIII and a selectable marker: in this case dihydrofolate reductase. Following selection on methotrexate, the cells were cultured in fermentors and the product harvested and tested.

However, it was clear that there were problems in production since the end product was two to three times less active than would be expected based on calculations of specific activity. The root of the problem appeared to be related to the size of the protein that resulted in inefficient translocation to the Golgi apparatus. Since these organelles are responsible for glycosylation, proteins failing to reach the Golgi apparatus were not glycosylated and therefore were inactive. The answer to the problem was provided by biochemical analysis of the protein and of patients with partial deletions of the gene. Analysis of data suggested that there were areas of the protein whose removal did not affect activity. Thus, genetic engineers made constructs in which these regions were deleted and then expressed the construct in CHO cells. The end product was smaller than the native product but was easier to make and was as active. There are still, however, some patients who raise antibodies to Factor VIII, both recombinant and natural. Efforts are currently focused on identifying the immunogenic regions and constructing a novel active molecule in which these regions are removed or masked.

4.3.6 Transgenic pigs and Factor VIII[47]

Factor VIII is now being produced in the milk of transgenic pigs.[48] The yields are said to be ten times greater than animal cell culture, the process cheaper and the product safe. However, fears of transmission of viruses from pigs to man have stopped the use of transgenic pig hearts, so is there a problem? We are now much more aware of xenozoonoses (the transmission of organisms from animal to man) and we are therefore legally obliged to ensure that any recombinant protein produced in pigs is safe.

4.4 All recombinants must be tested before they are given to humans

It is illegal and immoral to test new products directly on humans, and a series of cell culture experiments and animal tests must be carried out before a product may be tested on human subjects. There are, however, doubts about the validity of the tests. For example *in vitro* tests do not highlight problems of immune recognition of the product. Cultured cells are not normal and often have abnormal chromosome numbers; hence they may produce far more of a particular enzyme required to reduce toxicity of a product than normal cells do. Material designed for use in human systems may well fare badly in animal tests and those products deemed safe in animal models might not work in humans.

Once developed, the product must also be licensed by a government body to ensure that only validated products are used therapeutically. The type of information required is outlined in Tables 4.5 and 4.6.

Table 4.5   Tests required to ensure safety and efficacy

- *in vitro* determination of product activity and purity
- *in vivo* tests in animals to determine dose/response, adsorption, distribution, excretion and toxicity
- *in vivo* tests on healthy humans
- clinical studies: pilot and multi-centre

Table 4.6   Information required for licensing a recombinant protein

- nature of the gene
- restriction map of gene + vector
- DNA sequencing of the gene
- genetic stability results
- description of cells
- cell bank information
- source material of DNA
- production method
- fermentation method
- harvesting procedure
- purification regime
- characterisation of the product
- comparison with 'normal' material
- analytical methods
- validation process
- impurities detected
- batch analysis
- additive information
- control tests
- stability tests
- analytical tests including metabolism and bioavailability
- diluent and packaging analysis

The product is then authorised for use.[49] The time span between conception of a product and production may be as long as 10 years. The biotechnology company in that time must fund the product research without reward. Venture capitalists invest money in the companies at great risk, since few products make it to the marketplace. The investment, however, may reap great rewards for the shareholders since the total market for recombinant proteins was estimated to be $250 billion in 2003.

4.5 Why make recombinant proteins?

This is perhaps a question we should have addressed earlier. We have already talked about the lack of safety in human-derived material. Indeed, of 5000 haemophilia patients in the UK treated with blood-derived Factor VIII prior to HIV testing, 1200 are HIV positive and 207 have died. Patients treated with growth hormone and fertility hormones died of CJD. Patients are also at risk from hepatitis C which may lead to liver cancer, and recently Borna virus,[50] which is responsible for depression in humans, has been suggested as a new problem. The situation is now so uncertain with respect to CJD transmission that it has been recommended that plasma from the UK should not be used in the isolation of proteins for therapeutic use. However, we should contemplate whether other countries really have a safer product than does the UK. We know prion disease developed in individuals receiving cadaveric proteins from countries in which CJD has not been reported to be a problem.

So how safe is blood- and cadaveric-derived materials? Whether recombinant proteins are indeed safer than human-derived materials is a debatable point. Many of the cell hosts used could harbour viruses. Calf serum, if used as a growth additive, may also add risk. On a more practical note, quality control and assurance are easier to achieve with recombinant proteins than with blood-derived material – there should be no batch-to-batch variation and yields should be higher.

It is very important to remember that without protein therapy, haemophiliacs die. Natural protein products saved lives and this must be balanced against the numbers who were infected in the 1980s with HIV. Even if recombinant proteins caused deaths in some patients, provided that the majority gained benefit, then the therapy has been worthwhile. No therapy is completely safe and risk : benefit ratios should always be calculated prior to use.

4.6 Recombinant products

There is a plethora of products now made as recombinants. We have already discussed examples of hormones and clotting factors, but there is a growing market in recombinant vaccines, growth factors, antibodies and cytokines[51] Vaccines, antibodies and cytokines will be discussed separately in Chapters 5 and 6.

Table 4.7 Products available as recombinants[52]

- growth factors, e.g. erythropoietin, thrombopoietin, platelet-derived growth factor, bone morphogens, myotrophin
- cytokines, e.g. interferons and interleukins
- hormones, e.g. insulin, human growth hormone and atrial natriuretic peptide
- receptors, e.g. CD4
- fibrinolytics, e.g. tissue plasminogen activator and streptokinase
- vaccines, e.g. malaria and hepatitis B
- monoclonal antibodies, e.g. Herceptin
- clotting proteins, e.g. Factors VIII and IX, protein C and S
- enzymes, e.g. DNase and glucocerebrosidase
- haemoglobin

4.7 Generics

The patents of many of the early recombinants will soon come to an end and biotechnology companies worldwide will be able to make their own copies of these products, which are termed 'generics'. The issues, specifically quality control and testing, need to be addressed.[53,54]

Points to consider

Name six companies involved in the manufacture of recombinant proteins and list their key products

Why aren't all replacement proteins made as recombinants?

Why are recombinants so expensive?

Are recombinant products safe?

Notes

31 www.kazusa.or.jp/codon/

32 http://en.wikipedia.org/wiki/Insulin_recombinant_human

33 www.biocenter.helsinki.fi/biotechgs/media/toikkanen.pdf

34 www.nature.com/nbt/journal/v22/n11/pdf/nbt1028.pdf

35 www.mnstate.edu/provost/BaculovirusExp.pdf

36 http://uwadmnweb.uwyo.edu/NEWSLETTER/2003/october/caterpillar.htm

37 www.nature.com/nbt/journal/v22/n11/pdf/nbt1026.pdf

38 www.nature.com/cgi-taf/DynaPage.taf?file=/nbt/journal/v18/n11/full/nbt1100_1151.html

39 Powell, K. Barnyard biotech – lame duck or golden goose. *Nature Biotechnology*, 2003, **21**, 963–7

40 www.pnas.org/cgi/reprint/94/26/14677

41 Greenwell, P. Blood group antigens: molecules seeking a function? *Glycocon. J.*, 1997, **14**, 159–73

42 Jenkins, N. *et al.* Getting the glycosylation right: implications for the biotechnology industry. *Nat. Biotechnol.* 1996, **14**, 975–81

43 Berglund, B. and Wide, L. Erythropoietin concentrations and isoforms in urine of anonymous olympic athletes during the Nagano Olympic Games. *Scand. J. Med. Sci. in Sport*, 2002, **12**, 354–7

44 Fandrey, J.K. and Jelkman, W.E. Chemical structure, biotechnological production and clinical use of recombinant erythropoietin. *Z. Gesamte Inn. Med.*, 1992, **47**, 231–8

45 http://en.wikipedia.org/wiki/Factor_VIII

46 www.fda.gov/bbs/topics/NEWS/NEW00312.html

47 Paleyanda, R.K. *et al.* Transgenic pigs produce functional human factor VIII in milk. *Nat. Biotechnol.*, 1997, **15**, 971–5

48 www.ag.unr.edu/tittiger/bch405/notes/Gen_Engr_I/animals.pdf

49 www.mhra.gov.uk/

50 Takahashi, H. *et al.* Higher prevalence of Borna disease virus infection in blood donors living near thoroughbred horse farms. *Journal of Medical Virology*, 1998, **52**, 330–5

51 Walsh, G. Biopharmaceutical benchmarks. *Nat. Biotechnol.*, 2003, **21**, 865–70

52 http://en.wikipedia.org/wiki/List_of_recombinant_proteins

53 http://ndt.oxfordjournals.org/cgi/reprint/20/suppl_4/iv31

54 Schellekens, H. When biotech proteins go off-patent. *Trends Biotechnol.*, 2004, **22**, 406–10

5

Recombinant vaccines

Recombinant vaccines are essentially recombinant protein products that are made by incorporating genes, usually from pathogenic organisms, into other cells. On administration, the protein elicits an immune response and induces immunological memory so that when challenged by the protein the patient will be immune. In this chapter we will discuss the history of vaccines, their evolution and problems associated with their use, the rationale for the production of recombinant vaccines and the use of bioinformatics and proteomics in vaccine design.[55]

5.1 Vaccine history

There is evidence to suggest that the Chinese used vaccination as a preventive strategy against smallpox more than 3000 years ago. In 1000 BC Egyptians recorded variolation, the introduction of small amounts of pus from smallpox victims into the skin to prevent disease. The first written account describes a Buddhist nun around AD 1022 to 1063 grinding up scabs taken from an infected person into a powder, and then blowing it into the nostrils of a non-immune person. By the 1700s, this method was common practice in China, India and Turkey. In the late 1700s European physicians used this and other methods of variolation, but reported some 'devastating' results. Overall, 2% to 3% of people who were treated died of smallpox, but this practice decreased the number of fatalities due to

Molecular Therapeutics: 21st-century Medicine by Pamela Greenwell and Michelle McCulley.
© 2007 John Wiley & Sons, Ltd

smallpox to a tenth of the previous level. Members of the aristocracy used pus from smallpox victims as a vaccine and there was a thriving trade between Turkey and the rest of Europe.

Towards the end of the 18th century, Edward Jenner (1749–1823) noticed that milkmaids rarely caught smallpox and suggested that their exposure to cowpox protected them from smallpox. Experiments using attenuated cowpox as a vaccine were successful and the method was rapidly taken up. In order to test his theory, in 1796, Jenner took the fluid from a cowpox pustule on the hand of a dairymaid and inoculated an 8-year-old boy. In an experiment that would be considered unethical today, he exposed the boy six weeks later to smallpox, and the boy did not develop any symptoms. The term 'vaccine' was coined by Jenner from the Latin word 'vaca' which translates as 'cow'. By 1800 about 100,000 people had been vaccinated worldwide.

Indeed, in the early part of the next century, in some areas of Britain it was an offence to be unvaccinated. The 'modern' smallpox vaccine that was licensed in the USA by the Food and Drug Administration (FDA) was developed by Wyeth Laboratories and licensed under the name Dryvax from a weak strain of virus, the New York City Board of Health strain. In 1967 the World Health Organization (WHO) started a campaign for global eradication of smallpox – this was achieved within 10 years. The last endemic case of smallpox occurred in Somalia in 1977 and on 8 May 1980, the World Health Assembly declared the world free of smallpox. Despite eradication, stocks of smallpox virus are still held worldwide. There have been moves to destroy these, but there was fear that if one country kept stocks they could be used in 'germ warfare'. Hence, though no longer a natural threat, smallpox still exists albeit as a military threat.

The early pioneers of vaccines faced many problems, as illustrated in *War on Disease: a History of the Lister Institute*.[56] In 1889 the UK government could not see the value of investing in an Institute for Vaccinology, specifically for rabies, as treatment was available in Paris, only 8 hours from London. Indeed, the trip to Paris was said to be cost-effective: '£25 to cover travel and accommodation for two for two weeks which in total was £750 a year for those needing treatment'. This was considered a bargain compared to setting up an Institute. This history also relates the fate of a number of pioneers of vaccines: Ruffer and Plimmer working on diphtheria anti-toxins both developed the disease but recovered on administration of the anti-toxin.

In the 1890s Pasteur produced vaccines to anthrax and rabies and Behring used antibodies from patients who had successfully recovered from diphtheria as an anti-toxin. In the 1950s vaccines were developed for chicken pox and measles. More recently, in a bid to develop safer vaccines, recombinant and DNA vaccines have been developed.

5.2 Vaccines

Vaccines are usually made against whole cells or components of bacterial, protozoan and viral infections to stimulate *ACTIVE immunity*. This is our traditional concept of vaccinology. In this case an antigen is administered, an immune response evoked, memory cells are produced and protection is gained from further assault, e.g. childhood triple vaccine (measles, mumps and rubella). *PASSIVE immunity* is used if there is not enough time to induce active protective immunity. It is derived by the administration of preformed antibody to the causative organism or toxin, e.g. rabies, diphtheria toxin, anti-venom, human anti-rhesus. If a patient has been infected with rabies for example and has not yet been vaccinated we can administer a preformed antibody to clear infection. This is obviously important for diseases where a mass vaccination scheme would not be cost-effective, for example rabies which does not exist in the UK. Passive immunity provides short-term immunity only; a problem encountered with passive immunity is serum sickness if the antibody has been derived from animals.

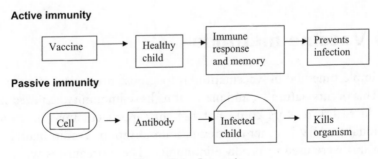

Figure 5.1   Immunity

The aim of a vaccine is to stimulate production of memory cells so that when the person is re-challenged with the organism they can rapidly produce an effective immune response and clear the organism before any damage may occur. In the humoral response, B-cells bind to released copies of antigenic proteins and then multiply. Many of the progeny secrete antibody molecules that during an infection would home in on the pathogen and mark it for destruction. Others become memory cells that will quell pathogen if it circulates outside cells. The display of antigenic protein fragments, or peptides, on the surface of inoculated cells (within grooves on MHC class I molecules) can trigger a cellular response. Binding to the antigenic complexes induces cytotoxic (killer) T-cells to multiply and kill the bound cells and others displaying those same peptides in the same way. Some activated cells will also become memory cells. To set the stage for B-cell activation, 'professional' antigen-presenting cells (APCs) must ingest antigen molecules, process and display the resulting peptides on MHC class II molecules. Helper T-cells, in turn, must recognise both the peptide complexes and 'co-stimulatory' molecules found only on the professional antigen presenters. Before the cytotoxic cells can respond to antigens on inoculated cells, APCs have to take up vaccine plasmids, synthesize the encoded antigens, and exhibit fragments of the antigens on MHC class I molecules along with co-stimulatory molecules. Then the killer T-cells must recognise those signals and also be hit by cytokines (this time of the Th1 type) from helper T-cells.

Vaccines are developed after rigorous testing for efficacy and safety; they must be of proper dosage to stimulate immune response and provide immunity for several years without inducing hypersensitivity or autoimmunity. Ideally they should be inexpensive, easy to store and administer and most importantly, must be safe.[57]

5.3 Vaccine methods

The simplest method of 'vaccination' is to expose the patient to the pathogen. This occurs naturally, and many of us have immunity acquired in this way. In reality doctors could not use this approach. Imagine exposing individuals to HIV in order to attempt immunisation. Traditionally, whole organisms were used to produce immunity. These organisms were killed, inactivated or weakened (attenuated). It would be unwise to attempt vaccination with a live organism since the effects could be devastating. However, it is important for the preparation to be immunogenic and many dead and inactivated organisms lose their power to induce an effective immune response.[58]

Table 5.1   Examples of vaccines in use

Vaccine	Preparation
BCG (TB)	Live attenuated
Measles	Live attenuated
Mumps	Live attenuated
Polio (Sabin) – oral	Live attenuated
Polio (Salk)	Inactivated
Yellow fever	Attenuated
Cholera	Dead
Influenza	Inactivated
Pertussis (whooping cough)	Killed
Plague	Formaldehyde killed
Rabies	Inactivated
Typhoid	Killed
Typhus	Killed

Most inactivated vaccines are considered to be dangerous since they may be re-activated to give live organisms.

5.4 Types of vaccine

Examples of vaccines produced by inactivating a pathogen are the Salk polio vaccine and the influenza vaccine. Typically the virus is grown in culture and then treated with formaldehyde so that it cannot reproduce in patients. The advantages of this method are that the risk of infection is low and the resulting vaccine can be used safely on the immunocompromised. The disadvantages are that as it cannot multiply the vaccine needs to contain a large number of virions to stimulate immunity, does not induce memory, must have boosters and must be injected, and is therefore costly.

An example of an attenuated vaccine is the Sabin polio vaccine. The pathogen is grown in conditions that make it less virulent; the advantage here is that the vaccine stimulates memory cells, and less virus needs to be injected as the virus can still multiply, and additionally the Sabin vaccine can be administered orally and so therefore is less expensive. One of the problems of attenuated vaccines is that in rare instances the virus reverts to virulent form and can cause disease; therefore this type of vaccine should not be given to the immunocompromised. There have been calls to administer the Salk vaccine first and then the Sabin to reduce the chances of live virions being passed in faeces – a problem seen in young children treated with the Sabin vaccine alone.[59,60] However, there are reports that the Sabin vaccine induces immunity in those exposed by proxy to infants who have been vaccinated and that this in turn increases herd-immunity.

5.5 The limitations of vaccine programmes

Vaccine programmes are expensive to develop since there is no guarantee of success. A good example is malaria. An important problem associated with protozoal diseases is that the organisms exist in many forms within the body and are therefore difficult to target. Additionally, they have developed sophisticated mechanisms to alter their cell surface proteins and sugars in order to evade the immune system. This in turn makes them poor candidates for vaccine development. The development of HIV vaccines has also proven to be difficult. Again the organism is highly mutable, hides within cells and coats itself with sugars which appear to the host to be 'self'. Additionally, we know that an antibody produced in the serum of an individual with the disease does not affect disease progression since almost all infected individuals have serum antibodies. Ideally mucosal responses are required to help prevent the disease from establishing itself in the host. Even if vaccines could be easily developed there are other problems to consider such as how the vaccines will be tested and how their efficacy will be assessed. Once developed and tested there are problems of delivering the vaccine to the individuals.

Vaccines are not cheap, but in fact can be cost-effective when compared to other alternatives. A good example is cholera infection: one option is vaccination, the other is to set up new clean water supplies. The latter seems like an attractive option, but in unstable countries if conflict breaks out the water supply will be damaged and therefore disease risk will reappear. On the other hand, once a population is vaccinated they are safe regardless of the quality of the water supply. Ideally of course we should use both strategies together. Education is also a problem in persuading people to come forward for vaccination. In the USA a survey found that only 30% of health workers had taken up the option to be vaccinated against hepatitis B. How can we make people take part in the programme? The problem is that the unvaccinated act as a reservoir of infection for the whole population. There are also problems with liability issues for the manufacturers. No vaccine can be 100% safe, so companies will need to insure themselves against claims for damage. In the UK there have been law suits against manufacturers of the MMR (measles, mumps and rubella) vaccine.[61] This followed newspaper reports that there was a link between this vaccine and autism.[62] There was an immediate withdrawal by parents of children from vaccine programmes which in turn has led to one of the worst outbreaks of measles since the vaccine was introduced. Similar problems have been seen with other vaccines following suggested links with disease. In the case of MMR, it is likely that children who responded badly to the vaccine would also have suffered from complications associated with the disease. Nevertheless it should be remembered that in the USA alone, prior to vaccination 300–700 deaths were reported every year due to measles, 20,000 children were born with birth defects due to maternal rubella and 1000 children died annually from mumps.

The vaccine against whooping cough has also presented the medical profession with problems. The disease is severe, causing death in some cases. However, the vaccine is not perceived to be safe. Even published documents agree that there are reactions to the vaccine. It is said that 40% of children have some reaction local to the site of injection – this is not unusual for a substance designed to stimulate the immune system. However, 40% also suffer from anorexia, vomiting, fever and crying, 20% have a reaction severe enough to warrant medical intervention and in 1 in 100,000 cases (0.001%) the child suffers from convulsions and hypotonia which may affect them for the rest of their lives. If the vaccine were proven to be 100% successful, these figures could be accepted but in fact the vaccine is only 80% effective: that is, 1 in 5 vaccinated children may contract the disease. Obviously there is need here for a new, safer and more effective vaccine. In the UK in 2006, there was a resurgence of whooping cough in young adults, which was due in part to the limited effectiveness of the vaccine.

Almost everyone in the UK has been vaccinated against tuberculosis using the BCG vaccine. The vaccine itself was first developed in 1908 and has not changed in 90 years. Its continued use may suggest that this is therefore a very effective product. Additionally, cases of TB have dropped dramatically during the last hundred years. However, this had little to do with the vaccination policy. TB was a disease of poverty, poor nutrition and poor hygiene. Improvements in the lifestyle and health of the population have been more effective in the drop in TB numbers than has the vaccine programme. However, TB is still a problem in many developing countries where conditions have not significantly improved and where vaccine programmes have not been freely available. Currently there are 8 million new cases of TB reported per year, with 3 million deaths, most of which are in the developing world. However, there has been an increase in the incidence of TB in immunocompromised patients, particularly those who are HIV-positive, in the developed world. More worryingly, these individuals are harbouring antibiotic-resistant strains of the disease. TB has never been an easy disease to treat, requiring intensive drug therapy over a period of 6 months. The disease is so feared that in some states in the USA sufferers thought unlikely to comply with the drug regime may be imprisoned for the duration of treatment.

The threat of antibiotic-resistant TB has forced scientists to look critically at the current vaccine provision. Worldwide, 70% of children have been vaccinated and 3 billion doses of vaccine have been used. This sounds reassuring; however, in a study in Malawi scientists showed that children vaccinated 8 years earlier retained no immunity to the disease. In the UK the vaccine seems effective, but the risk of meeting an infected individual is very low. In India, where TB is endemic, the vaccine offers little or no protection. There is obviously much concern about the problem and many groups are trying to make more effective vaccines.[63]

5.6 The role of the WHO

Despite some of the problems listed above, immunisation has played an important role in eradicating disease. The immunisation campaign carried out by the World Health Organization (WHO) from 1967 to 1977 effectively eliminated natural occurrence of smallpox. In the early 1960s, the disease threatened 60% of the world's population and killed 25% of those infected. The WHO has stated that eradication of poliomyelitis is within reach; however, there have been problems recently with isolated outbreaks and non-compliance in vaccine programmes. Since the launch of the WHO-sponsored Global Polio Eradication Initiative in 1988, infections have fallen by 99%, and it is estimated that 5 million people have escaped paralysis. Between 1999 and 2003, deaths from measles decreased worldwide by almost 40%, and some regions have targets for eliminating the disease. In 14 of 57 high-risk countries, maternal and neonatal tetanus have been eliminated.

5.7 Problems specific to developing countries

Since drugs are prohibitively expensive, vaccines offer a better option in developing countries where prevention is not only better than cure but is cheaper and more effective than cure. The WHO and other UN bodies have worked hard to ensure that vaccines are available to as many children as possible.[64] However, there are many diseases that occur only in the developing world for which vaccines are not being developed. Parasitic diseases such as schistomiasis, filariasis and helminth infection cause some of the most debilitating diseases. In total more than 1000 million people suffer from such infections. Parasitic infections are also a severe problem in Third World countries. It is estimated that hundreds of millions suffer from malaria. Indeed, malaria is said to have killed or debilitated more soldiers in wars in the Middle and Far East than the conflicts themselves. Vaccine trials are in progress, initiated not by the WHO or UN but by the US Army. Leishmaniasis is another protozoal infection that again affects hundreds of millions in the developing world. Diseases such as meningitis are also a major cause of death in parts of Africa. The picture here is further complicated by the lack of resources to diagnose and treat the disease: thus many fatalities are seen. The WHO reported over 700,000 cases of meningococcal disease in the period 1988–97; this is known to be a significant underestimate. The number of meningitis-associated deaths was 100,000 during this period, and in 1996 alone more than 180,000 cases occurred. The death rate was about 10% even when healthcare was available. In sub-Saharan Africa, death rates were as high as 30% with at least 75,000 children having sustained central nervous system injury following administration of 'oily' chloramphenicol, an antibiotic emulsion that is not recommended in developed countries.[65] Last, but by no means least, HIV 1 and 2 continue to be major problems in the Third World where heterosexual and materno-fetal/neonate transmission occurs.

In the UK and the USA drug therapies are available for the treatment of AIDS but the costs are in excess of £50,000 per patient per year. In countries such as Uganda, there is no chance that such regimes will be available since the health budget is only $6 per person per year. Zidovudine (AZT) which is known to prevent materno-fetal transmission, is also unavailable in many parts of the developing world. In others, such as South Africa, although drug companies offered reduced prices or even free drugs the governments refused them on the grounds that they would need to test all pregnant women and did not have the infrastructure to do so. Although new cheaper treatments are being developed, vaccines are really the only option for such diseases. Once developed they are a cheap and effective method of disease prevention. However, development costs are large, protozoal and viral diseases are difficult targets, and we must address issues of safety.[66]

5.8 Economics and logistics of vaccinology

Many current vaccine programmes require multiple dosing. If patients live in inaccessible areas, they will be required to attend remote clinics to which the vaccines must be delivered. In hot climates it is important that the heat and humidity do not degrade the vaccines. This is a major problem since refrigeration is not available in many areas of the world. Samples may be shipped to the receiving country under ideal conditions but may then be left in hot customs houses before transportation to the clinics. Some developing countries make their own vaccines; this divorces vaccination from profit and helps to reduce the cost of vaccines. However, syringes and needles are still a problem. They are expensive and therefore frequently reused. Without adequate sterilisation diseases such as HIV may then be spread from person to person during the vaccination programme. There are also no vaccines for some of the biggest killers in developing countries such as respiratory infections (pneumococcus and haemophilus), diarrhoea (rotavirus) and malaria.

On 2 March 2000, the US National Institutes of Health (NIH) reported[67] that HIV had killed 2.6 million people worldwide in 1998, tuberculosis had been responsible for 1.5 million deaths and malaria for 1.1 million. The reported expenditures on potential vaccines were $250 million, $6.5 million and $25 million respectively – hardly a fair distribution! President Clinton announced a 40% greater budget for TB research and a 10% increase in budget for malaria vaccine research in 2000.

Despite the problems we face, vaccines save 1.5 million children a year. This can be compared to the impact of nutrition improvements and clean water. The latter costs $1260 per life saved whereas vaccination costs $10 per child. UNICEF aimed to immunise 90% of babies against common infectious agents such as measles by the year 2000. In 1990 they reported a 70% uptake of vaccines compared to 1974 when the figure was 5%. UN organisations fund vaccinology: $15 million is spent on research and $100 million purchasing vaccines, refrigeration, sterilisers, needles, syringes, fuel and vehicles. With so little of the budget allocated for research it is probably not surprising that for many diseases vaccines remain a wish rather than a reality.

In 2007 the United Nations General Assembly Special Session (UNGASS) pledged that by 2010 there would be full immunisation of children under 1 year of age at 90% coverage nationally with at least 80% coverage in every district or equivalent administrative unit.[68] In 2002, WHO estimated that 1.4 million of deaths among children under 5 years were due to diseases that could have been prevented by routine vaccination. This represents 14% of global total mortality in children under 5 years old. WHO targets are: diphtheria, *Haemophilus influenzae* (Hib), hepatitis B, measles, meningitis, mumps, neonatal tetanus, pertussis, poliomyelitis, rubella, tetanus, tuberculosis and yellow fever. Some diseases, including poliomyelitis, measles and maternal and neonatal tetanus, have specific goals for eradication or elimination.

A study showed that a one-week immunisation initiative against measles in Kenya in 2002[69] in which 12.8 million children were vaccinated should prevent 3,850,000 cases of measles and 125,000 deaths and result in a saving of US$12 million of the health budget over the next ten years. In the USA, it is estimated that one dollar invested in a vaccine dose saves US$2 to US$27 in healthcare expenses. In the 1990s, vaccines for tuberculosis, polio, diphtheria, tetanus, pertussis and measles cost about US$1 per child. The addition of vaccines for hepatitis B and HiB raised the cost in the developing world to US$7–13 per child excluding administration and injection equipment; when administration was included, costs rose to US$20–40 per child.

International initiatives such as the Expanded Programme on Immunization and the Global Alliance for Vaccines and Immunization (GAVI)[70] are providing the impetus, funding and technical support to help increase the number of vaccines provided and immunisation coverage. The proposed WHO–UNICEF Global Immunisation Vision and Strategies which is projected to run from 2006–2015, will expand vaccination coverage, and allow the setting up of logistical systems. More recently vaccinology has received help from the Bill and Melinda Gates Foundation whose remit, with respect to global healthcare is:

to encourage the development of lifesaving medical advances and to help ensure they reach the people who are disproportionately affected. Funding is focussed in two main areas:

- Access to existing vaccines, drugs, and other tools to fight diseases common in developing countries
- Research to develop health solutions that are effective, affordable, and practical

The fund was set up with an endowment: $33.4 billion. Since inception a total of $13.6 billion has been committed to grants and in 2006 grant payments were $1.56 billion. To date, for example, the fund has made donations of $1.5 billion to the GAVI alliance and $258 million to the Malaria Vaccine Initiative. It is also supporting a number of HIV programmes worldwide.[71]

Clearly there is a need to develop cheap, effective and heat-stable vaccines for use worldwide. This may be achieved by research into employing molecular tools in the development of recombinant, gene and DNA vaccines.

5.9 Recombinant vaccines

Many of the problems associated with vaccines could be solved by making vaccines by isolating proteins from the organism or by making recombinant protein versions of these proteins. Genetic engineering could also be harnessed to provide more heat-stable versions of the vaccine for use in developing countries and edible vaccines are also being produced. Effective vaccines have been produced for a number of infections using purified proteins as vaccine. These include the vaccines for diphtheria and tetanus where the purified toxin is treated with formaldehyde before use, hepatitis where the cell surface antigen from disease carriers may be used, and meningitis and pneumonia where purified polysaccharides form the basis of the vaccine. However, isolation of proteins and sugars is time-consuming and may prove difficult. Therefore there has been a drive to produce vaccines as validated recombinant proteins.

There are three main approaches to the delivery of recombinant proteins to the patient: to produce pure protein and allow the patient to inject, inhale or ingest the protein; to utilise the vaccinia virus where the gene for the vaccine is cloned into the vaccinia genome and administered to the patient who expresses the recombinant protein and elicits an immune response; DNA vaccines.

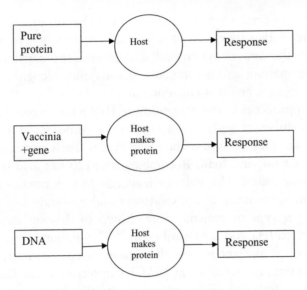

Figure 5.2   Vaccine strategies

5.9.1 Simple recombinant protein vaccines

These are the recombinant protein vaccines in contrast to those produced in vaccinia. The first recombinant vaccine was the recombinant hepatitis B surface antigen vaccine which was produced in 1982. Originally the protein was produced in yeast cells, but later developments showed that mammalian cells were more effective in production of this heavily glycosylated protein. All vaccine programmes involving hepatitis B are now using the recombinant version of the vaccine.

Recombinant vaccines encompass subunit vaccines, subunit vaccines with adjuvants and conjugate vaccines; examples include DPT (diptheria, pertussis and tetanus). They contain a purified antigen rather than a whole organism and adjuvants such as alum to enable slow release and so provide more immune stimulation. Conjugates increase immunogenicity, for example HiB conjugated with dipthoid toxin. They are semi-synthetic vaccines made of polysaccharide antigens from microorganisms attached to protein carrier molecules. The carrier protein is recognised by macrophages and T-cells, thus enhancing immunity. Conjugate vaccines enable us to combat the relative inability of young children to mount an effective immune response against encapsulated bacteria, especially HiB, *Neisseria meningitidis* (Nm) and *Streptococcus pneumoniae* (Sp). A booster dose later in life creates a robust and rapid antibody response, indicating the existence of immunologic memory in primed children. Antibodies induced by conjugate vaccines are functionally active. The advantage of recombinant vaccines is that they are not infectious; on the negative side they do not stimulate the immune system well and the antigens may not retain their native conformation, so immunoglobulins may not recognise antigen on the real pathogen when it is encountered.

Vaccine approaches to the prevention of HIV have involved a variety of targets. Recombinant gp160[72] resulted in a response in HIV-infected patients in that they showed a change in clonal B-cell expression but this had no impact on viral load. Recombinant gp120 has also been investigated but is ineffective. The melanoma-specific MAGE tumour antigen has been used in approaches to the treatment and prevention of melanoma. Fifty-seven per cent of patients were shown to develop an increase in MAGE-directed IgG and IgM antibodies.[73] The immune system does not usually recognise the antigen on the tumour itself; however, antibodies raised in response to recombinant MAGE can recognise the same antigens on the tumour itself and elicit immune recognition of the tumour as non-self. Recombinant cholera vaccine is used as an oral preparation and recombinant TB vaccines are currently in trial.

Recombinant vaccines have been trialled for HIV infection. However, HIV is a continuing problem as the virus 'hides' in cells and is difficult for the immune system to detect, it is highly mutable, its reverse transcriptase has no proof-reading capacity and it replicates rapidly. We know a great deal about the structure and genetics of the virus but we cannot formulate either a cure or a vaccine. People infected with HIV produce antibodies but do not clear the infection, thus a vaccine which simply induces antibodies is unlikely to be successful.

There is worldwide diversity of HIV variants: certain HIV 1 strains are common in the USA and Europe whilst HIV 2 and other strains of HIV 1 are more common in Africa, India and Thailand. Many of the vaccine strategies are targeted towards strains common to the USA and Europe. In 2004 there were more than 30 clinical trials of candidate HIV vaccines reported: most were small phase I trials to assess safety or generate preliminary immunogenicity data; none appeared be safe or efficient. Strategies include: inactivated viral particles, viral antigens, DNA vaccines, mucosal vaccines, recombinant vectors and infection with attenuated HIV.[74] The latter was carried out despite animal data suggesting that the virus could be reactivated. Current vaccines in trials include the recombinant monomeric HIV-1 envelope gp120 vaccine (AIDSVAX; VaxGen Inc, Brisbane, CA, USA); however, the results from phase III trials (AIDSVAX B/B) have been disappointing (i.e. 3.8% efficacy). There has also been a phase III AIDSVAX B/E trial in Thailand, the outcome of which was unfortunately disappointing. A third trial where the vaccine was combined with a booster component, attenuated canary pox vector, is also currently in phase III (ALVAC vCP1521, Aventis Pasteur).[75] The IAVI database gives a comprehensive list of AIDS vaccines in human trials.[76]

5.9.2 Gene vaccines: the vaccinia virus approach

Gene vaccines can be made by inserting the gene of interest into a vector, such as recombinant vaccinia, injecting into the patient, allowing to replicate, express gene and produce large amounts of antigen *in situ*. The most successful demonstration of this approach has been in the eradication of rabies in Western Europe. The *Rabies glycoprotein* gene was inserted into the vaccinia genome and the vaccine was placed in food. Since the trial was to attempt to eradicate rabies from the feral fox population the vaccine was loaded into chicken heads which were dispersed over a large area. The vaccine was shown to give 18-month protection against rabies in adult animals and is cost effective when compared to culling. The success of this vaccine, and the eradication of rabies from France and Belgium led to the UK withdrawing its strict quarantine regulations which had been in force for nearly a century.

5.9.3 DNA vaccines

There is now a move to make vaccines by injecting the patient directly with DNA encoding the sequence of the recombinant to be produced. In this case, instead of putting the gene into an expression system, making the protein and delivering it to the patient we are giving the DNA directly to the patient and allowing them to make the recombinant protein *in vivo*. The major advantages are that the mechanism is efficient and that DNA can be transported at ambient temperature without degradation. Hence these gene vaccines could play an important role in countries where lack of refrigeration is currently limiting the vaccination programmes.

DNA vaccines use 'naked' DNA that is directly injected into muscle or coated onto gold particles and shot into the skin by pressurised gas ('gene gun') or via liposomes in creams. This has been tested in life-threatening situations rather than in routine vaccinology as there are concerns about long-term safety. By isolating a harmless piece of a pathogen's DNA and injecting it into the body, the host makes the protein and the immune system develops a response, both humoral and cellular, against a particular disease even though the body was never exposed to it. Whereas gene therapy tries to work in spite of the immune system, DNA vaccines harness the immune system's instinct to search out and destroy non-self proteins. DNA vaccines hold special promise against diseases too complex or dangerous for traditional vaccinology.

DNA vaccines elicit protective immunity primarily by activating humoral responses, which attack pathogens outside cells, and cellular immunity, which eliminates cells that are colonised. Immunity is achieved when long-lasting 'memory' cells are generated. On entry of a DNA vaccine into a targeted cell, such as muscle, the cell produces antigens.

5.9.3.1 DNA vaccination for malaria

Currently there are at least 35 candidate malaria vaccines in clinical development, many of which are or soon to be in clinical trials. Vaccine strategies target different aspects of the complex life cycle of malaria to prevent infection, prevent disease or block transmission. Three of the vaccine candidates studied in target paediatric populations in field settings are:

- RTS, S/AS02A: safe, highly immunogenic, immunity in ~50%, efficacy wanes with time, at phase 2

- MVA-ME TRAP: good safety, most advanced, at phase 2

- MSP-1/AS02A: phase 1.[77,78]

5.9.4 Edible vaccines from transgenic plants

Vaccines have been made successfully in a range of plants including bananas and potatoes. In the banana, the recombinant vaccine is expressed in the fruit. Individuals eating the fruit would receive enough of the vaccine to raise an immune response. It has been suggested that the fruit should not only express the vaccine, but also a gene encoding a coloured pigment (blue has been suggested), to differentiate the plant from normal banana plants. In theory seeds or young plants could be sent to each village in, for example Africa, where they would be nurtured and the fruit used in vaccination programmes. In countries where bananas will not grow, the same strategy could be applied to potatoes or tomatoes. The product would be cheap, stable and freely available to each community. Edible vaccines could in theory overcome refrigeration and sterility problems as well as being cheap and so are ideally suited to developing countries. The disadvantages associated with edible vaccines are that the expression levels of the transgene are low and that there is variability in immunogenicity and stability. There are also serious issues that relate to genetically modified crops.[79] Nevertheless there are edible vaccines[80] in development targeted against rabies,[81] TB[82] and diarrhoeal diseases.[83]

5.10 Rational design: bioinformatics and proteomics

Recombinant protein production also allows us to genetically engineer the product. In 1997 Rappuoli and his colleagues described what they termed the 'first truly engineered vaccine'. The group was working on vaccines for pertussis and employed their knowledge of the structure of the bacterial toxin to generate a safer but no less effective product. Essentially, they used site-directed mutagenesis to change the protein structure of the toxin; specifically they were interested in changes which reduced toxicity of the toxin but would still elicit an immune response which would be directed at both the native toxin and the recombinant.[84]

The vaccine works well and is said to have an efficacy 84% better than current vaccines and a potency of 10–20 times greater. It took the group from 1988 until 1998 to produce a marketable product. This highlights the investment of both time and money in bringing a new product to fruition. Currently this group is making use of the genome data available on microbes to mine their genomes for likely vaccine targets.[85] This approach can reduce vaccine development time significantly and is being used by many researchers.

In the next decade we should start to see new and more effective vaccines entering the market. Most will have been synthesised using molecular technologies and many will have been designed using the available genome databases and proteomics tools. In essence, proteomics tools can be used to determine which genes of the organism genome are expressed on, for example, the cell surface. To ensure a good immune response, the target proteins need to be substantially different in the organism compared to the host. This type of '*in silico*' analysis reduces the number of candidate proteins that need to be tested *in vivo*, saving both time and money.

5.11 Other interesting areas for vaccine development

Vaccination could play a role in removing proteins or cells from the human body and is being investigated for a wide range of diseases. We will look at the role of vaccines in cancer in subsequent chapters. In Alzheimer's disease the strategy is to reduce protein tangles and the formation of lesions in the brain. For atherosclerosis, in the fight against heart disease, vaccines could be used to generate antibodies to cholesteryl ester transfer protein (CETP) which may increase the ratio of high- to low-density lipoproteins. Vaccines have also been suggested for drug addiction, specifically cocaine and nicotine, as a contraceptive strategy and even for obesity.

5.12 Conclusion

Vaccines are a proven cost-effective way of preventing disease and their development, with the continued support of the WHO and other UN bodies, will continue to have massive impact in many developed and developing countries.

Points to consider

Why has it been difficult to make vaccines against malaria and HIV?

What are the benefits and disadvantages of recombinant vaccines compared to traditional vaccines?

A close friend of yours has recently become a parent, and states that the baby will not be immunised because immunisations are dangerous and anyway no one in this country gets 'those diseases' any more. What is your response to your friend and upon what evidence do you base your advice?

Investigate the controversy over vaccine (Gardisil) against human Papilloma virus (HPV). This is a sexually transmitted infection and we are aiming to vaccinate all 10–12 year olds! Would you let your children be vaccinated?

Notes

55 www.vaccineinformation.org/video/index.asp

56 Chick, H. *et al. (1971) War on Disease: a History of the Lister Institute*. The Trinity Press, London

57 www.niaid.nih.gov/publications/vaccine/pdf/undvacc.pdf

58 www.niaid.nih.gov/factsheets/evolution_vaccines.htm

59 Shulman, L.M. *et al*. Oral poliovaccine: will it help eradicate polio or cause the next epidemic? *Isr. Med. Assoc. J.*, 2006, 8, 312–15

60 Yoshida, H. *et al*. Prevalence of vaccine-derived polioviruses in the environment. *J. Gen. Virol.*, 2002, 83, 1107–11

61 www.parliament.uk/documents/upload/POSTpn219.pdf

62 www.cdc.gov/nip/vacsafe/concerns/autism/autism-mmr.htm

63 www.pubmedcentral.gov/articlerender.fcgi?tool=pubmed&pubmedid=15176980

64 www.who.int/topics/en/

65 Robbins, J.B. *et al*. Meningococcal meningitis in sub-Saharan Africa: the case for mass and routine vaccination with available polysaccharide vaccines. *Bull. World Health Organ*; 2003, 81, 745–50

66 http://www.cdc.gov/od/science/iso/about_iso.htm

67 Birmingham, K. Industry outlines its perspective on new Third-World vaccine development *Nature Medicine*, 2000, 6, 723–4

68 www.who.int/immunization_monitoring/en/

69 www.redcross.org/press/intl/in_pr/020614measles.html

70 www.gavialliance.org/

71 www.gatesfoundation.org/GlobalHealth/

72 Rollman, E. The rationale behind a vaccine based on multiple HIV antigens. *Microbes and Infection*, 2005, 7, 1414–23

73 Ollila, D.W. *et al*. Overview of melanoma vaccines: active specific immunotherapy for melanoma patients. *Seminars in Surgical Oncology*, 1998. 14, 328–36

74 www.iavireport.org/trialsdb/searchresults1.asp?list=vaccine&vt=.&ts=ongoing

75 Garber, D.A. *et al*. Prospects for an AIDS vaccine: three big questions, no easy answers. *Lancet Infect. Dis.*, 2004, 4, 397–413

76 www.iavireport.org/trialsdb/searchresults1.asp?list=vaccine&vt=.&ts=ongoing

77 Ballou, W.R. *et al*. Update on the clinical development of candidate malaria vaccines. *Am. J. Trop. Med. Hyg.*, 2004, 71, 239–47

78 Enosse, S. *et al*. RTS,S/AS02A malaria vaccine does not induce parasite CSP T-cell epitope selection and reduces multiplicity of infection. *PLoS Clin. Trials.*, 2006, 1, e5

79 Streatfield, S.J. Regulatory issues for plant-made pharmaceuticals and vaccines. *Expert Rev. Vaccines*, 2005, 4, 591–601

80 Korban, S.S. *et al.* Foods as production and delivery vehicles for human vaccines. *Journal of the American College of Nutrition*, 2002, **21**, 212S–217S

81 Ashraf, S. *et al.* High level expression of surface glycoprotein of rabies virus in tobacco leaves and its immunoprotective activity in mice. *J. Biotechnol.*, 2005, **119**, 1–14

82 Rigano, M.M. *et al.* Oral immunogenicity of a plant-made, subunit, tuberculosis vaccine. *Vaccine*, 2006, **30**, 691–5

83 Tacket, C.O. *et al.* Plant-derived vaccines against diarrheal diseases. *Vaccine*, 2005, **23**, 1866–9

84 Pizza, M. *et al.* Genetic detoxification of bacterial toxins. *Methods Mol. Med.*, 2003, **87**, 133–52

85 Serruto, D. and Rappuoli, R. Post-genomic vaccine development. *FEBS Lett.*, 2006, **580**, 2985–92

6

Therapeutic antibodies and immunotherapy

This chapter is not designed to investigate the nature of the immune response but is focused on the use of genetically manipulated agents in the treatment of disease. It would therefore be difficult to ignore antibodies and cytokines since they are produced as recombinant proteins and comprise about 25% of the bio-drugs produced. All of the constraints we have previously discussed in the production of recombinant proteins also apply to antibodies and cytokines.

6.1 Monoclonal antibodies

The idea of using antibodies as 'magic bullets' to treat diseases such as cancer has been around for more than 25 years. However, despite the high hopes of the original researchers they were not able to harness antibodies as therapeutic tools. In the main, the early failures were due to the fact that proteins produced in mice are immunogenic in humans. Therefore, after the first injection, the product was rapidly cleared from the circulation.

Prior to the production of monoclonal antibodies all antibodies were derived from the sera of animals. These were polyclonal antibodies. Polyclonals comprise a mixture of antibodies directed at a range of different epitopes within the target protein. Such specificity is useful when mutations may have occurred in the protein since an epitope-specific antibody may

Molecular Therapeutics: 21st-century Medicine by Pamela Greenwell and Michelle McCulley.
© 2007 John Wiley & Sons, Ltd

not react if its target epitope has been changed. Using a range of antibodies or polyclonal serum prevents such an occurrence. However, polyclonal antibodies have limited value as therapeutic tools, since, if they were derived from animal sources, they will be recognised as non-self by the immune system and rapidly cleared. Nevertheless some antibodies for therapeutic use, for example antibodies to diphtheria toxin, were made in the horse. Patients, however, became ill with serum sickness – essentially an immune reaction. Monoclonal antibodies were traditionally made in mice. The individual antibody-producing cells are fused to myeloma cells to form immortal hybridomas each of which will express an antibody with a single specificity. Specificity for a single epitope may be desirable but many companies blend a range of monoclonals to produce a cocktail of antibodies recognising different parts of the same molecule.

Both polyclonal and monoclonal antibodies could be used to deliver passive immunity in cases of rabies or tetanus. Traditionally products from the sera of previously infected humans or animals were used as a source of antibody. However, apart from the reactions to the antibody itself, batch variation was a problem. Additionally, material derived from humans must now be treated with caution, as, unless screened, it could be a source of infection.

Table 6.1   Polyclonal antibodies used in therapy

Antibody	Source
Hepatitis B	Human
Measles Ig (immunoglobulin)	Human
Rabies Ig	Human
CMV Ig	Human
Tetanus Ig	Human
Botulism anti-toxin	Horse
Diphtheria anti-toxin	Horse
Snake venom anti-toxin	Horse

6.2 Monoclonal production

Monoclonal antibodies in theory offer a continuous supply of a validated and specific product but the specificity may in itself be a problem. Monoclonal antibodies were traditionally made in mice using the method shown in Figure 6.1.

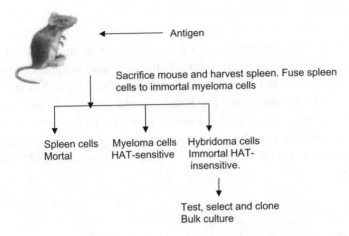

Antigen

Sacrifice mouse and harvest spleen. Fuse spleen
cells to immortal myeloma cells

Spleen cells   Myeloma cells   Hybridoma cells
Mortal   HAT-sensitive   Immortal HAT-
                                   insensitive.

Test, select and clone
Bulk culture

Figure 6.1 Steps in antibody production

The immunisation regime will ensure that the animal is producing the appropriate antibody prior to sacrifice and testing must be carried out to check the specificity and sensitivity of the antibody raised. The mouse is then killed and the spleen harvested and the cells disaggregated by rubbing through a sieve. The cells are then resuspended in medium, washed and mixed with a known amount of myeloma cells. The addition of polyethylene glycol (PEG) mediates cell fusion. The fused cells are grown on a selective HAT (hypoxanthine, thymidine, aminopterin) medium. Cells with the enzyme hypoxanthine phospho-ribosyl transferase (HPRT) grow in the presence of HAT by utilisation of the alternative pathway to nucleic acid synthesis. Myeloma cells are HAT-sensitive since they lack HPRT, spleen cells are HAT-insensitive but die after only a few weeks in culture. The hybridoma cells gain immortality from the myeloma cells and HPRT from the spleen cells and hence after a few weeks of growth in selective medium only hybridomas remain viable. The hybridomas are grown at low density in 96-well plates and the supernatant tested for antibody activity.

Positive cells are cloned by diluting to single cell density and re-growing and then tested. When the clonality has been assured the cells are then grown in bulk culture for use. In theory every B-cell and hence each derived hybridoma will produce a different antibody and provided that we select for the antibody we require we can produce large amounts of a very specific reagent. There are, however, problems associated with the production of monoclonal antibodies, some of which we have already addressed in preceding chapters. Since antibodies are produced in mouse cells they may carry the immunogenic Galα1-3Gal epitope which in turn would result in rapid clearance of the antibody from the system. The amino acid composition of immunoglobulins is also different in mice and humans and therefore this may cause additional problems. For therapeutic use, the antibody molecules produced should pass easily into cells and organelles. Essentially, the take-home message from those researchers using traditional monoclonals was that although a mouse monoclonal could be used once the human anti-mouse response (HAMR) precluded further use.

6.3 Therapeutic monoclonal antibodies[86]

In order to be a useful therapeutic tool an antibody must be human-compatible and retain its half-life in the human host, that is it must be sufficiently glycosylated. The specificity must be such that we can use the antibody to target specific cells by virtue of the proteins they express to simply detect the proteins in solution. Finally the antibody should be as small as possible in order to allow penetration. Since, for the reasons outlined, murine monoclonal antibodies are not suitable for therapy a number of alternative approaches have been investigated. These include the production of human monoclonal antibodies, engineered humanised antibodies, antibodies made in transgenic mice and antibody fragments made in bacteria. None is ideal and all have problems associated with them.

6.3.1 Human monoclonals

In theory these would provide an ideal solution producing a guaranteed human-compatible product. However, we cannot ethically immunise humans and remove their spleens to allow us to produce antibodies. Some individuals have already been exposed to infectious agents or proteins and therefore produce a range of 'natural' antibodies and it is possible to harvest peripheral blood B-cells, immortalise these with Epstein–Barr virus (EBV) and create an antibody-producing cell line. However, cells which produce antibody are difficult to immortalise and cells which are easy to immortalise do not produce antibody. Additionally, viral contamination may be a problem when using reagents derived from humans and in the current climate such procedures are frowned upon.

One success story involves the production of a human monoclonal antibody recognising the Rhesus D antigen (RhD).[87] RhD isolated from human blood had been used for many years in the treatment of Rh-negative women carrying Rh-positive babies. On production of the human monoclonal, the serum reagent was replaced and used in therapy. Women who have carried a Rh-positive child but who themselves are Rh-negative may be exposed to the fetal red cells at birth, during amniocentesis or CVS (chorionic villus sampling). Fetal cells in the maternal circulation then prime the immune system. If the mother then carries a second Rh-positive child her preformed antigens may recognise the fetus as non-self and attempt to destroy it, causing haemolytic disease of the newborn. Treatments for severe cases involve inter-uterine transfusion of the fetus, replacing their Rh-positive blood with Rh-negative. Currently all Rh-negative women at risk of carrying an Rh-positive child are treated with anti-RhD antibodies to ensure that all fetal cells in the maternal circulation are cleared before they can prime the immune system. Anti-RhD is now produced as a recombinant antibody.[88]

6.3.2 Humanised antibodies

For many antigens there are no individuals available as a source of 'natural' antibodies and hence human monoclonals cannot be made. In this case genetic engineers can use recombinant DNA techniques to engineer the production of an antibody whose binding site is derived from a murine monoclonal *but* whose backbone is derived from a human Ig. The antibody engineering used to involve protein engineering but nowadays usually involves genetic engineering and the polymerase chain reaction (PCR).[89] For much of the work in which we are interested, the crucial part of the antibody molecule is the combining site, that is the area of the immunoglobulin that binds to the epitope. Other parts of the molecule are important for effector function and the induction of aspects of the immune response.

In order to generate humanised antibody we need to understand the nature of the combining site for the recognition of a specific antigen. We

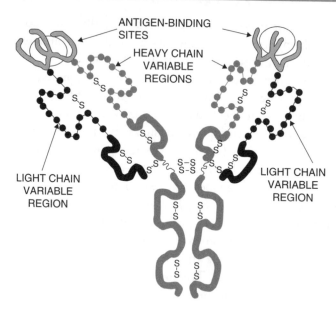

Figure 6.2   A schematic of an antibody

can then design gene constructs in which the majority of the sequence is that of a human immunoglobulin but the area defining the combining site is that derived from the mouse antibody. Such a small area of murine protein is undetectable as non-self and these antibodies are well tolerated in therapy. The gene constructs can be expressed in transgenic animals, cell culture or even in bacteria provided no effector function is required since this is mediated by the glycosylation.

In some cases, for example in cancer therapy,[90] there is a requirement for small antibody fragments to penetrate tumours or cells. Additionally, if the antibody is merely to be used as a transporter of other molecules, size constraints are an important feature. Hence, many companies have become involved in the production of antibody fragments for use in therapy. Essentially these may be as small as the combining region of the antibody or single-chain antibody fragments (SFv). There have been numerous developments in the characterisation, isolation and production of small antibody fragments. These are discussed more fully in other chapters. Suffice it to say here that a technique called 'phage display'[91] is used to select antibody fragments that bind to the antigen of choice. Basically, a λphage library is made from cells exposed to the immunising antigens. The phage is engineered to express its proteins on its surface. If each phage infects one bacterium, we can use a screening technique using labelled antigen to select clones expressing the antibody of choice. Once isolated, the gene can then be manipulated and the experiment repeated using fragments of the gene as construct, selecting for the smallest fragments that still bind to the antigen. Once selected, the clones can be grown in bulk to produce therapeutic amounts of product. It is also possible to make libraries from unstimulated B-cells, each of which will have a novel immunoglobulin gene re-arrangement. In this way, virtually any immunoglobulin a human could produce can be made.

6.4 Transgenic monoclonals

Human monoclonals can now be made in transgenic mice. There is a mouse called the Severe Combined Immune Deficiency (SCID) mouse[92] which lacks a functional immune system. Researchers have replaced the bone marrow of these mice with human stem cells. Hence, on immunisation they will produce antibodies of choice, which are characteristically human.[93] Other researchers have expressed human immunoglobulin genes in transgenic mice. The mouse immunoglobulin gene is replaced by human immunoglobulin gene at the one-cell stage of the egg. The animal then will express the human version of the immunoglobulin gene and be stimulated to produce essentially human antibodies. Transgenic plants can also be used to produce antibodies and antibody fragments. The plants of choice are *Nicotiana* and tomato. A humanised monoclonal antibody that protects against genital herpes (HSV-2) has been synthesised in soy bean.[94,95,96,97]

6.5 The uses of monoclonal antibodies in therapy

Therapeutic strategies with monoclonal antibodies can be divided into two clear groups: *ex vivo* and *in vivo*. *Ex vivo* suggests that material is removed from the patient, treated outside the body and then returned to the patient. In this case the problems of reaction to the antibodies used are minimal. The second strategy involving *in vivo* work has a strong requirement for human compatibility of the antibody. Examples of *ex vivo* strategies include purging bone marrow either of cells that will cause rejection (in the case of allografts – grafts from other individuals) or of cancer cells (in the case of autografts – grafts from oneself). Antibodies may be used to remove toxins *ex vivo*, passing for example human blood over immobilised antibody to remove small metabolites. Recently there has been much interest in the isolation of haemopoeitic stem cells using antibodies. These stem cells may be used to repopulate marrow in patients receiving autografts or as a vehicle for gene therapy. *In vivo*, antibodies can be used to deliver cytotoxic drugs, toxins, radioactivity and heavy metals specifically to their site of action, minimising damage to normal cells. In gene therapy, antibodies have been suggested as a mechanism to target DNA to cells.

6.6 Specific examples of therapeutic strategies

Strategies using unconjugated antibodies or antibodies linked to agents designed to damage the target cells have been developed. Each of these will be discussed using relevant examples.

6.6.1 Unconjugated antibodies – treatment of kidney rejection

Since there is a shortage of organs available for transplant, many researchers have focused on preventing organ rejection, therefore making better use of the organs available. A number of companies produce monoclonal antibodies designed to remove or inactivate the T-lymphocytes involved in the rejection process. The antibody produced by the company Cantab binds to CD45-positive cells and mediates their destruction. Used *ex vivo*, the antibody is used to treat the organ prior to transplant. In clinical trials, treated kidneys gave increased organ survival *in vivo* and the response was dose-dependent. There was no evidence of side effects since the manipulation was carried out outside the body. The OKT 3 antibody works in a similar manner, but is used *in vivo*. The antibody is used to destroy mature T-cells that could mount an attack on the organ and prevent interleukin 2 binding and rejection. However, side effects were reported with this antibody. The cost of the antibody is also a problem at £2000 per 50 mg. Nevertheless, increased success of transplant is itself cost-effective.[98]

6.6.2 HIV treatment

In HIV research monoclonal antibodies are used to combat secondary infections. Since the patients become immunocompromised, they are prone to severe infections from opportunistic agents. The antibody MSL 109 has been produced to combat the cytomegaloviral (CMV) infections seen in these patients. This frequently presents as CMV retinitis. Another antibody to hepatitis B is being used to reduce the viral load in HIV-positive patients who have developed hepatitis B. Additionally, a number of anti-viral antibodies are in trial that target HIV itself. These include antibodies to the viral reverse transcriptase.

6.6.3 Cancer therapy and monoclonal antibodies

Monoclonal antibodies have been seen as a breakthrough in cancer therapy. However, there are problems with their use. These include the targeting of the antibody to the tumour. There are few if any molecules that are cancer-cell-specific and many markers of cancer are merely up-regulated or mutated forms of natural cellular products. Additionally, cancers are heterogeneous with respect to antigen production and therefore an antibody may not recognise every cancer cell. Sadly, removal of all cancer cells is a prerequisite for a cure. One of the first tasks is therefore the recognition of a target molecule on the cancer cell.[99,100]

6.6.4 Herceptin in breast cancer therapy

Despite problems with identifying targets, antibodies are used widely in cancer therapies. One of the most exciting was approved by the FDA for patient use in 1999, having successfully completed clinical trials. This antibody, Herceptin (trastuzumab), is directed at HER2. Originally the antibody was termed MSK. Trastuzumab is the generic name given to the humanised antibody composed of the binding region recognising HER2 combined with a human immunoglobulin G to decrease immunogenicity and enhance the recruitment of immune effector cells. The antibody recognises the growth factor product of the HER2 oncogene (Her2/neu) which is expressed in abundance in many breast cancers. In the early trials,[101] 46 patients with breast cancer were investigated and all received a weekly dose of the antibody until the cancers were seen to progress. Very few side effects were seen and of the 46 patients 1 had complete remission, 4 had 50% shrinkage of the tumour mass, 2 had 25% shrinkage and 14 stabilised. We need to view these findings in the light of the fact that the patients chosen were terminally ill women who had tried every available therapy. One would not expect miracle cures of such sick patients, but the response was very good. Later trials suggested that the treatment was more effective on less sick women. More recently, studies have shown that patients treated with a combination of antibody and chemotherapy, specifically paclitaxel or doxorubicin and cyclophosphamide, have an advantage (25.4 months vs. 21 months survival). However, Herceptin plus doxorubicin causes cardiac toxicity in some patients. Many women with breast cancer unfortunately cannot benefit from this therapy since they shed the HER2 protein into their blood stream. This effectively inactivates the antibody before it reaches the cancer. Patients must be tested pre-prescription for their HER2 status, that is, whether they over-express the protein and it is shed into the blood stream. In this way, only those who would benefit from the drug would be given it. In the UK, there have been problems of prescription with the so-called 'postcode prescribing' of the drug in some areas of the country and not others.[102] However, some emotive newspaper reports suggesting patients are being denied the drug reflect a lack of understanding of the effectiveness of this drug in a subset of cancer patients.

6.6.5 Treatment of multi-drug-resistant cancer cells

Another approach to the treatment of cancer is to target the P-glycoprotein drug pump. P-glycoprotein is present in all cells and is used to pump drugs from the cell into the surrounding fluid, thus preventing cell damage. Over-expression of this molecule is seen in many cancers, and effectively the pump then removes from the cells the cytotoxic drugs used in therapy. Therefore, high levels of drugs would be required to achieve an effect, but this would damage normal cells whose drug pump was not over-expressed. The working hypothesis is that if we could target the glycoprotein with antibodies, we could effectively turn off the pump and render the cells susceptible to drugs. In early cell culture trials[103] this approach was shown to convert drug-resistant cells into drug-sensitive cells. In animal studies using rats injected with drug-resistant ovarian cancer cells, those treated with 10 µg of antibody had complete tumour loss whereas those treated with 1–10 µg showed slowing of tumour progression. Untreated animals died. However, one of the major criticisms of this type of animal model is that the tumour is experimental and is not given time to metastasise. Hence the researchers allowed the tumours to develop in some animals before administration of the antibody and the same results were obtained with 4 or 5 animals apparently tumour-free. The mice all regressed within 42 days of the treatment. It is difficult to judge how effective such treatment would be in humans.

6.6.6 Anti-endotoxin antibodies

Antibodies to endotoxins have been developed by many companies, although the products are most notable for their lack of success. Sepsis is a problem associated with gram-negative organisms that can cause death of the affected patient. The active molecule is lipid A. The antibody developed by one biotechnology company is a good example of the type of results obtained. The E5 antibody was developed to bind to lipid A and was used to treat patients with acute renal failure and non-shock sepsis morbidities. Results from the preliminary trials, treating patients with sepsis originating from gram-negative infections, were excellent. Compared to the untreated control there was a greater than twofold increase in the number of patients cured. However, as the trials moved on, patients with gram-positive infection and normal individuals were used as controls. Sadly, the side effects of the antibody killed 4 or 5 of the control group. Within hours the shares in the company had fallen and the company came close to failing because of this single product and its unpredicted results.

6.6.7 Conjugated antibodies

Conjugated antibodies have been developed for use in transporting other molecules to specific cells.[104] The antibody is simply used as a transporter and we make use of its specificity to target cells or molecules for delivery of the cargo. One of the first uses of this technology was *in vivo* pre-surgery imaging. This involved the conjugation of heavy metals to antibodies that would recognise cancer cells. The antibody was given prior to surgery and the location of the antibody determined and deemed to outline the affected area to be removed. In this case murine antibodies were used since patients are only ever exposed to the antibody on one occasion.[105]

6.6.8 Delivery of radionuclides to leukaemia and lymphoma patients

There have been numerous reports of successful delivery of radionuclides to tumours using antibodies as targeting agents.[106] Preliminary studies showed that in lymphoma patients the best strategy was delivery of high-level rather than low-level radioactivity. The patients needed to have marrow support but the response was 100% in the high-level group compared to 10–40% in the low-level. The reverse was seen in the treatment of leukaemia. In a series of experiments in Seattle, an antibody cocktail to the cell markers CD37 and CD20 labelled with radioactive iodine was delivered to patients with B-cell lymphoma. The antibody specifically recognises B-cells and was used to treat B-cell lymphomas, acute myeloid (AML) and acute lymphoid (ALL) leukaemias. Of the 20 patients treated, 17 appeared to enter complete remission with a median remission time of 9 months. Nine patients entered remission and were disease-free for more than 4.5 years. The major problems associated with this technique relate to the myelotoxicity of the radionuclides themselves. This limits the effective dose that can be delivered. The radioactivity has only a short range of about 2 mm, which is useful in that the radiation does not affect surrounding tissue but has the disadvantage that we can only treat small tumours. The uptake of the complex by the tumour was only 1% and there were problems associated with free antibody complex in the system. If the radionuclide is removed from the antibody it is free to damage cells indiscriminately. In fact high levels of the radionuclide appeared in the kidneys of patients following treatment and this could result in renal failure.

6.6.9 Drug delivery

Antibody-mediated drug delivery has also been attempted. In this case the idea is that toxic substances can be taken safely to cells and we can deliver much higher doses of drugs to cancer cells specifically. Drugs have been encapsulated in liposomes for delivery but the liposomes cannot easily move out of the blood stream. It is possible to take inactive drugs to cancer cells by these methods, relying on the enzymes within the cells to activate these 'prodrugs' to yield active drugs.

6.6.10 Toxin delivery

The delivery of toxins such as ricin, endotoxins, mitogellin and RNase by antibodies is attractive. These molecules are so dangerous that they must be directed to the tissue to be killed in such a way as to avoid damage to normal tissue, for example, ricin delivery to cancer cells. Ricin prevents ribosome assembly resulting in cessation of protein translation; thus it is clearly very toxic. In the case of ricin, there is an important drawback: ricin is cleared by the liver. The interaction and recognition is through carbohydrate residues; thus if ricin is synthesised in E. coli no sugars are added and the resultant toxin does not locate to the liver. Researchers have also used endotoxins, mitogellin and RNase as conjugation partners. The problem is how to prevent toxins killing cells on the way to the cancer cells. One solution is to conjugate the antibody and the toxin using disulphide bonds which remain reduced in the circulation. On entry to cells, bonds are broken and the toxin is freed and acts on the cell. Photolinkers can also be used to couple the toxin to the antibody: in this case, the toxin is activated by lasers. Instead of toxins we can use antibodies to deliver enzymes such as Xanthine oxidase which breaks down DNA.

6.6.11 Bispecific antibodies

These are antibodies with two combining sites. This enables them to be used to target antigens on two cell types and then bring them together to allow interaction. They may also be used to bring together a cell surface antigen and a soluble antigen, for example, to bring a cell to metabolise a specific protein at the site of protein deposition.[107]

6.7 Other recombinant proteins used in immunotherapy

6.7.1 Cytokines

The cytokines[108] are a large family of proteins produced in small amounts by one cell activating another. In a manner similar to hormones and neurotransmitters they allow cell–cell communication. However, whereas hormones are released by specific organs and tissues into the blood stream and neurotransmitters are released by neuronal cells, cytokines may be produced and released by a range of cell types. Cytokines play an important role in both innate and adaptive immune responses and are involved in immunological and inflammatory responses. Individual cytokines are produced as part of a response and often are responsible for the development of a cascade of events. Thus, by manipulating one cytokine, the natural balance of the response may be altered in such a way that, instead of effecting a cure, we simply trigger another problem. Sadly, many of the cytokines are toxic or mediate severe reactions when used in therapeutic strategies. Research into the nature and function of individual cytokines has advanced rapidly since the relevant genes were cloned and recombinant proteins produced. Thus, production of recombinant cytokines has led to the availability of marketable quantities that could not have been isolated by any standard biochemical procedure.

6.7.1.1 Cytokines of therapeutic interest

The types of cytokines of commercial interest are the interferons, interleukins, colony-stimulating factor and tumour necrosis factor. As previously stated, our interest is in the fact that these were some of the most widely produced recombinant proteins. Whether they have a therapeutic niche is debatable. Interferon α was one of the first recombinant proteins, having been produced for trial in 1986. Originally it was marketed as 'a universal panacea'; however, its usefulness is limited by the side effects produced. Since that time, a range of cytokines have been produced and entered clinical trials. Few, however, have found proven clinical application. Cytokines have been used in the treatment of a wide range of disorders and many companies are involved in marketing. However, despite the 'hype' they really have little use.

Table 6.2   Diseases that have been treated with cytokines

- cancer
- AIDS
- rheumatoid arthritis
- multiple sclerosis
- asthma
- allergies
- infections

Table 6.3   Cytokine products of the biotechnology industry

Products	Company	Disease
IL-2	Cetus	AIDS, renal cell carcinoma
		Karposi sarcoma
		Cancer
	Amgen	Cancer
	Hoffmann-La Roche	Cancer
IL-1α	Immunex	prevention of bone marrow suppression following chemotherapy and radiotherapy
IL-1β	Immunex	melanoma
	Syntex	immunotherapy, wound healing
IL-3	Hoechst-Roussel	bone marrow failure
	Immunex	platelet deficiency, BMT, stem cell transplants
IL-4	Schering-Plough	immunodeficiency
	Immunex	cancer therapy

6.7.1.2 Interleukins in trials and in use

Many of the large biotechnology companies market cytokines. Of those trialled the most successful would seem to be interferon β, which is used to treat patients with multiple sclerosis. Although not a cure this cytokine appears to reduce the number of relapses. However, many local NHS trusts were reluctant to prescribe this drug because of its cost. The National Institute for Clinical Excellence (NICE) published a review in which it stated that NHS money should not be used to prescribe interferon β and called for a review in November 2006.[109] Roferon (IFNα) is being marketed for cancer treatment and Actimmune for leukaemia treatment. There have been some limited successes with specific neoplastic disorders.

Table 6.4   Interferons in trials

Product	Company	Application
IFNγ Actimmune	Genentech	chronic granulomatous disease
IFNα Alferon N	Interferon Sciences	genital warts
IFNα Intron A	Schering-Plough	hairy cell leukaemia, genital warts, Karposi sarcoma, hepatitis C
IFNγ Actimmune	Genentech	small cell lung cancer, dermatitis, infection renal cancer, asthma, allergies
IFNβ Betaseron	Berlex	multiple sclerosis, cancer
IFNα Roferon	Hoffmann-La Roche	colorectal cancer, hepatitis, CML, AIDS

6.7.1.3 TNF: a cautionary tale

A note of caution should be made about the use of cytokines in treatment. Tumour necrosis factor (TNF) kills tumour cells and has been used in cancer treatment trials and in cases of rheumatoid arthritis. However, some patients developed septic shock-like symptoms. Doctors then realised that TNF is stimulated by bacterial lipopolysaccharides, a direct cause of septic shock. The TNF was the cause of the symptoms and instead of benefiting patients it actually made then unwell. TNF also causes wasting in some cancer and AIDS patients. So what did the biotechnology companies do with the TNF they made? They raised antibodies, recognising that anti-TNF may play a role in preventing sepsis and wasting in cancer! Indeed, some work in animals suggests that antibodies to TNFα may help prevent transplant rejection. Therefore antibodies to TNF seem a better therapy than TNFα. The problem is that TNF is part of a cascade and any interference with that cascade can cause catastrophic problems.

6.7.1.4 Adverse effects of cytokines

The adverse reactions to cytokines are well documented. They are essentially toxic, causing problems from trivial flu-like symptoms to severe fevers, inflammation and septic shock. Indeed, in early trials of cytokines on terminally ill cancer patients, a number withdrew from the trials because of the side effects. At present, analysis of the nature and function of the interleukins, interferons and TNF suggests that some recombinants will be used in therapy whereas other recombinant proteins may be used to raise antibodies to effectively shut down the cytokine-stimulated reactions.

6.7.2 Colony-stimulating factors and growth factors

These form another group of cytokines that are of particular interest to biotechnology companies. The factors stimulate growth and differentiation of cells from an immature stem cell to a specific and fully functional specialised cell. Since the advent of stem cell culture, the ability to produce cells of a particular lineage for a patient has been an exciting option. The first of these colony-stimulating factors isolated stimulated production of specific haemopoeitic cells *in vitro*. The factors were named according to the final cell produced, e.g. granulocyte stimulation factor (GSF). It is clear that these factors may be useful both to the biotechnology industry and to provide therapeutic options for patients.

Table 6.5   Characteristics of blood-cell-related growth factors

- stimulate production of specific cells types in culture
- produced by a variety of cell types
- little sequence homology between the factors isolated (G-CSF, M-CSF, etc.)
- receptors of these factors are glycoproteins embedded in cell membranes with ligand binding site
- factor binding to its receptor induces intracellular signalling events
- many cloned and expressed in *E. coli*
- administration of specific CSF promotes specific white cell differentiation and maturation *in vivo*
- used to treat patients susceptible to infection by raising circulating macrophage levels
- used to stimulate immune cell production in AIDS patients and cancer patients post-chemotherapy

Table 6.6   CSFs in use[110]

Product	Company	Application
Leukine (GM-CSF)	Bayer Healthcare	Use in 338,000 oncology patients in USA in 1992–2005 to increase neutrophil recovery and reduce infections
Neupogen (G-CSF)	Amgen	Approved in 1991 by FDA. Used in AIDS, leukaemia, aplastic anaemia and post-accidental radiation exposure
Leukomax (GM-CSF)	Sandoz Schering-Plough Genetics Inst	In use for treatment of low blood-cell counts and post-myelosuppression
Macrolin (M-CSF)	Chiron	In use for treatment of HIV patients

Points to consider

List the recombinant products used in immunotherapy.
Outline the problems associated with immunotherapy.
For what diseases is immunotherapy a useful tool?

Notes

86 http://users.path.ox.ac.uk/~scobbold/tig/new1/mabth.html
87 Avent, N.D. Recombinant technology in transfusion medicine. *Current Pharmaceutical Biotechnology*, 2000, **1**, 117–35
88 www.bloodjournal.org/cgi/reprint/103/11/4028
89 Presta, L.G. Selection, design, and engineering of therapeutic antibodies. *J Allergy Clin. Immunol.*, 2005, **116**, 731–6
90 Schaedel, O. and Reiter, Y. Antibodies and their fragments as anti-cancer agents. *Curr. Pharm. Des.*, 2006, **12**, 363–78
91 Russel, M. et al. Introduction to phage biology and phage display. www.oup.co.uk/pdf/0-19-963873-X.pdf
92 www.criver.com/research_models_and_services/research_models/RM_DS_SCIDMouseW.pdf
93 Weiner, L.M. Fully human therapeutic monoclonal antibodies. *J. Immunother.*, 2006, **29**, 1–9
94 Green, L.L. Antibody engineering via genetic engineering of the mouse: XenoMouse strains are a vehicle for the facile generation of therapeutic human monoclonal antibodies. *J. Immunol. Methods.*, 1999, 23111–23
95 Hwang, W.Y. and Foote, J. Immunogenicity of engineered antibodies. *Methods*, 2005, **36**, 3–10
96 Ko, K. and Koprowski, H. Plant biopharming of monoclonal antibodies. *Virus Res.*, 2005, **111**, 93–100
97 http://www.geocities.com/plantvaccines/futureprospects.htm
98 Ahssan, N. et al. Single dose intra-operative OKT-3 induction. www.chfpatients.com/tx/okt3_insert.pdf
99 Houshmand, P. and Zlotnik, A. Targeting tumor cells. *Curr. Opin. Cell Biol.*, 2003, **15**, 640–4
100 www.blackwell-synergy.com/doi/abs/10.1111/j.1745-7254.2005.00008.x
101 Mokbel, K. and Hassanally, D. From HER2 to herceptin. *Curr. Med. Res. Opin.*, 2001, **17**, 51–9
102 For reviews on targeting HER2, EGFR and VEGF in breast cancer, see www.medscape.com/viewprogram/2208_pnt
103 Pearson, J.W. et al. Reversal of drug resistance in a human colon cancer xenograft expressing MDRI complementary DNA by *in vivo* administration

of MRK-16 monoclonal antibody. *J. Natl. Cancer Inst.*, 1991, **83**, 1386–91

104 www.dekker.com/sdek/abstract~db=enc~content=a713491771~words

105 Borjesson, P.K. *et al.* Performance of immuno-positron emission tomography with zirconium-89-labeled chimeric monoclonal antibody U36 in the detection of lymph node metastases in head and neck cancer patients. *Clin. Cancer Res.*, 2006, **12**, 2133–40

106 http://users.path.ox.ac.uk/~seminars/Halelibrary/Paper%208.PDF

107 http://www.aapspharmaceutica.com/search/view.asp?ID=48070

108 www.cytok.com/

109 www.nice.org.uk/page.aspx?o=TA32

110 In *Nature Biotechnology*, **21**, 8 August 2003, there was a list of recombinant products, companies and patents

7

Transgenic animals

A transgenic animal is any animal whose genome contains some DNA from another organism. This is usually recombinant DNA and the resultant animal is therefore 'genetically engineered'. We need to be careful with this definition since many of us have 'rogue' DNA such as retroviruses incorporated into our genome, so we could also be loosely defined as transgenic!

Transgenic animals are made using germline gene manipulation: this allows us to introduce genes into egg cells in such a way that they can then be transmitted to future generations, allowing the creation of breeding stocks. The first transgenic mice were produced in the 1980s, with subsequent development of a range of transgenic animals including rabbits, sheep, pigs and fish for various research purposes. However, the majority of transgenic animals are still mice.

7.1 Why do we want to engineer the genomes of animals?

Making a transgenic animal is a costly and lengthy exercise and requires a special animal licence, so why do we make them? There are a number of legitimate reasons for carrying out work involving transgenic animals, for example in the production of disease models. Germline gene therapy –

Molecular Therapeutics: 21st-century Medicine by Pamela Greenwell and Michelle McCulley.
© 2007 John Wiley & Sons, Ltd

that is, manipulation of the genes in a fertilised egg such that any changes may be inherited – can be employed to target and 'knock out' a specific gene in order to look at the consequences to the animal. This is useful in cases where we are unsure of the role of a particular protein product in the pathogenesis of a disease and may be useful in order to generate an animal model for human disease. A good example of this is cystic fibrosis. There is no natural animal model for the disease and so transgenic mice[111,112] have been constructed which carry the mutant *CFTR* gene, so that new therapies that cannot be tested on humans can be tested on the transgenic animal. This, in theory, would also allow the toxicity of new drugs to be established without causing harm to humans. Indeed, such animal trials form part of the legislation for new drug products. Similarly, transgenic animals can be produced to model cancer and for studies of host–organism interactions in infection. In these cases scientists are particularly interested in susceptibility genes and their interactions. However, these models, apart from the transgene, are still mice and have different physiology, biochemistry, chromosome numbers and gene arrangements from humans, and their usefulness should still be questioned.

Genes can also be 'knocked in' to create a transgenic animal; in this case genes are inserted at a specific site known to produce high gene-expression levels. As discussed in preceding chapters, therapeutic antibodies are also produced in transgenic animals, both in the SCID mouse which itself lacks a functional immune system and in 'normal' mice. Transgenic animals have been made which contain gene constructs that will allow them to produce recombinant proteins in their milk. As this is not commercially viable with transgenic mice, transgenic goats, sheep, pigs and cows have been made. Animals, particularly pigs, are also being designed to produce human-compatible organs for transplant, and current research is trying to produce hearts for transplantation into humans. In the light of incidents such as BSE, some people question the safety of using products derived from animals. The main area of concern is the threat of xenozoonoses: infections crossing the species barrier.

In some countries the production of transgenic animals has been banned since the techniques involved use germline gene therapy. Animal rights activists argue that to produce and breed from animals manipulated to suffer from disease is cruel and unwarranted. In Great Britain in 2005 the numbers of transgenic animals used increased by 5% over the preceding year to 957,451. In contrast, the total number of procedures on animals in all areas of research increased by 1% to

2,896,198 over the same period. Procedures involving transgenic animals now account for 33% of all procedures conducted on animals in Great Britain.[113]

7.2 Experimental procedure

In order to produce a transgenic animal successfully we need to be able to artificially insert foreign DNA into the chromosomes of certain animal cells. The animal cells need to be totipotent, which means they have the capacity to differentiate into all the different cells of an adult animal. The types of cells used are fertilised oocytes, the cells of early-stage embryos and embryonic stem cells (ES). By making these genetically modified cells contribute to the development of a whole animal we thus make a transgenic animal.

Essentially, transgenic animals can be made in two distinct ways.

1. Straightforward germline manipulation through the random or targeted integration of exogenous DNA into chromosomal DNA of a fertilised oocyte.

2. Integration of DNA at post-zygotic level using embryonic stem cells. This method makes partially transgenic animals, chimeras, which are then mated. The resultant transgenic animals are seen only in the second generation.

7.2.1 Method 1: Germline manipulation

This method has been used to produce a variety of transgenic animals. Prior to making the transgenic animal it is necessary to make a transgene expression cassette and evaluate the expression of the transgene in cell culture. If this is successful then the transgene can be used to make a transgenic animal. The female animal is given drugs to induce super-ovulation and then she may either be mated naturally and after 24 hours the fertilised eggs collected by a method known as 'lavage', or the unfertilised eggs can be harvested and fertilised with sperm *in vitro*; the latter method is far more popular. The fertilised oocytes are then recovered; there are two visible pronuclei, the larger female and the smaller male.

About 2 pl (10^{-12} litre) of the desired DNA construct is microinjected through the zona pellucida into the male pronucleus. The transgene then randomly integrates into the animal's chromosomal DNA at a single site. The oocyte is re-implanted into a pseudo-pregnant surrogate mother and allowed to develop. At this stage a 'surrogate' animal is used as the original mouse cannot be used because the super-ovulation hormones interfere with normal implantation of the fetus. The surrogate must have been mated to a vasectomised male prior to implantation, because in mice a vaginal plug must be broken before necessary hormones are produced that allow implantation and maintenance of pregnancy. It is important that the surrogate does not become pregnant naturally as this would complicate the experiment.

When the pups are born, DNA is extracted from the tail tips (removed) and tested to determine whether they are indeed transgenic; this is normally done using DNA amplification (PCR) to detect the presence of the transgene sequence. Transgenic animals are then isolated and tested for expression of the mRNA and whether the protein of interest is correctly localised. The useful animals are then used to set up a breeding programme. In the case of disease models it is normally necessary to wait until maturity in order to observe pathology in the animal model. When transgenic animals are being made to produce a recombinant protein, expression of that protein is almost always targeted to milk. In those cases, it is necessary to wait until the animal is sexually mature and producing milk before success can be assessed. Males are not useful here except in a breeding programme.

7.2.1.1 Problems with this method

The success rate for production of a transgenic animal by this method is estimated as being only 10–25% efficient in mice and 1–5% in larger mammals. The main problems are transfer and insertion of the DNA transgene and problems associated with *in vitro* fertilisation (IVF). IVF is relatively efficient in mice, but not in larger mammals. In humans, for example, the success rate of IVF pregnancies is estimated at approximately 20%. Larger animals are less efficient with respect to implantation, and large numbers are wasted. Another problem is that random integration of the transgene into a gene encoding a key functional protein can result in insertional mutagenesis, whereby the key protein is disrupted and can kill or disable some pups. Additionally it has been found that a number of transgenic animals produced using this method are infertile and hence useless for future breeding programmes.

The time and effort to produce useful transgenic animals is immense. The larger animals have long gestation periods and mature slowly. It is financially draining to keep animals that may not be useful. If only 1% of the animals are transgenic all the others must be slaughtered, because they are genetically modified and so cannot be used as foodstuff. The second method, described below, was developed in response to a need for a more efficient method that ensures each experiment yields a transgenic animal with the appropriate gene incorporated at a single chosen site in the genome.

7.2.2 Method 2: Using embryonic stem cells

Scientists began to look at using embryonic stem cells as an alternative vehicle for transgenic production. Original experiments showed that embryonic stem cells, grown in culture, could be fused to blastulas produced by IVF after removal of the zona pellucida. The experiments[114] used blastulas from a white-fur mouse and embryonic stem cells (ES cells) from a black-fur mouse. The results were exciting: some pups were black, some white and others were chimeras with both black and white fur in their coats. Some of the chimeras passed on white-coat genes to their offspring whereas others passed on black. This suggested that embryonic stem cells could take on functions of the egg cells and that some of the offspring had reproductive tissues that had the genetic characteristics of the embryonic cell lines! The next approach was to introduce genes into embryonic stem cells to see if this could be a way of producing transgenic animals.

The first step in this process is to transfer the foreign DNA (transgene) into ES cells. Mouse ES cells are derived from 3.5-day post-coitum embryos and arise from the inner cell mass of the blastocyst. The ES cells are cultured *in vitro* and retain the potential to contribute to all mouse tissues when injected back into the host blastocyst and re-implanted. The developing embryo is then a chimera: it has two populations of cells derived from different zygotes, those of the blastocyst and those of the implanted ES cells. If the two cell populations are derived from mice with different coat colours then it is very easy to detect chimeric offspring. A chimeric mouse would be only partially transgenic; it is possible to derive a fully transgenic mouse by screening the offspring of matings between chimeras and mice from which the ES cells had been derived.

These early experiments were very important as they demonstrated that embryonic stem cells grow in culture, we can add in genes, select individual cells and grow clones, to determine exactly where the DNA has incorporated and test for the desired protein product. For example, the desired gene could be ligated to a marker gene such as the *neo* gene, enabling positive selection for transformant cells through growth on antibiotic-containing media. This would mean that only useful stem cells would then be fused with the blastula cells to produce the first-generation chimeras, a proportion of which will have reproductive tissue derived from the cell line that will be passed to their offspring. The chimera will express two sets of genetic information but the gonads will be derived from either the egg cells or the ES cells.

Thus, in order to make a transgenic animal using the ES cell method we first need to make a gene construct to infect ES cells in culture. The cells are grown in 96-well plates at low cell density, about 1 cell per well. Cells are selected that have the DNA incorporated at the appropriate locus by analysis of a proportion of the cloned cells. The selected cells that have the gene construct incorporated are fused to blastula cells from which the zona pellucida has been removed. The manipulated blastula is implanted in a host mother. When born, the offspring are mated to a normal animal. The DNA, RNA and protein expression are then tested in the offspring and if deemed to be successful a breeding programme is set up.

The major advantage of this system is that it enables us to be sure that the DNA is located in the appropriate locus before implantation. However, two problems still occur. First, many of the offspring are useless, as their reproductive tissue is derived from the blastula and not the stem cells and, since the first animals produced are chimeras, we will have to wait for a generation before we get true transgenic animals. This is not a particular problem with mice as they mature rapidly and their gestation period is short. However, in larger mammals there is a significant time delay between production of a chimera and production of a true transgenic. Additionally, since the blastula must still be transferred to the surrogate mother, there still remain the problems of implantation and successful pregnancy. Embryonic stem cells cannot be microinjected with DNA since we would need to manipulate thousands of cells; therefore viral vectors must be used to transfer the DNA construct into the embryonic cell line and with this the problems associated with using viral vectors that will be discussed in later chapters.

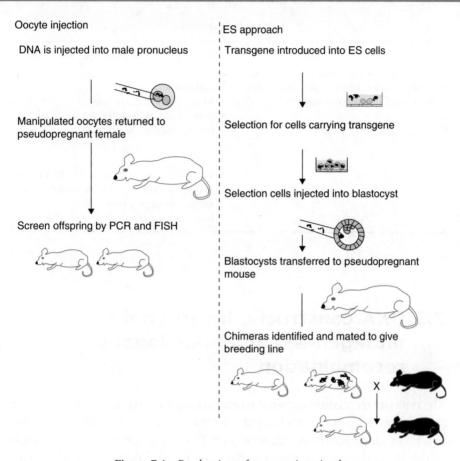

Figure 7.1 Production of transgenic animals

Table 7.1 Comparison of methods used to produce transgenics

Method 1: Germline manipulation	Method 2: Using embryonic stem cells
Method • Purified DNA injected into nucleus of single cell	**Method** • ES cells infected with construct in culture
Disadvantages • Site of insertion of DNA cannot be controlled • Inefficient DNA transfer	**Advantages** • Cells with construct in correct place selected and fused with blastula, then returned to host mother • Only cells with the DNA stably incorporated in the right locus are used
Uses • Used when animals are 'bioreactors'	

Table 7.1   Continued

Method 1: Germline manipulation	Method 2: Using embryonic stem cells
• Model disease phenotypes that result from gain of function mutation	**Disadvantages** • Many offspring are useless • Two generation times are required before results are known **Uses** • Best suited to large-scale production of transgenic animals and to sophisticated genetic manipulation such as homologous recombination

7.3 DNA constructs, insertional mutagenesis and homologous recombination

The type of DNA construct used when making a transgenic is an important consideration. Experiments have shown that linear DNA is taken up about five times more effectively than circular DNA (e.g. plasmids). The length of the DNA does not appear to affect the uptake although longer pieces of DNA are more difficult to handle than small ones. In reality the linear DNA forms concatomers (long strands of DNA all joined together), which integrate at a single site in the genome. Some researchers have used viral vectors to deliver DNA; although this does not appear to be more efficient than the traditional microinjection technique it requires less skill and takes less time. Other research has described the uptake of whole yeast artificial chromosomes (YAC) into mouse eggs and subsequent replication and gene expression. However, whole chromosomes are very difficult to handle and may be easily degraded *in vitro*.

The DNA construct in its most basic form will comprise the desired transgene to be inserted into the animal. We can rely on the host animal system to provide promoters and enhancers, but to ensure gene expression the DNA constructs should also contain these elements. This inevitably makes gene constructs much larger. If we are interested in expressing a gene in a particular organ, such as the mammary tissue, we also need to ensure that the construct contains tissue-specific expression signals. Studies of genes with tissue-specific activity have shown that there are sequences that bind activation factors produced only in the organ in which the gene is expressed. Thus, the gene is inactive in all other organs. Many researchers have studied 'upstream' sequences of genes whose expression is tissue-specific and have identified DNA binding sequences. If the correct sequence is incorporated into the construct then expression can be directed. This may be of real importance in disease models, for example if we model sickle cell disease we would like globin production confined to haemopoeitic cells. If we produce a clotting factor in a transgenic animal, expression in milk would yield product whereas expression in blood could kill the transgenic animal.

7.4 Use of inducible and tissue-specific promoters

For many applications it is desirable to have the transgene expressed under the control of a tissue-specific promoter or one that is inducible. In some cases coupled regulatory elements can confer position-independent and tissue-specific expression. Inducible promoters induce expression of the newly integrated gene. A good example is the tetracycline-regulated inducible expression of a transgene in transgenic mice whereby the expression of a reporter gene such as luciferase can be controlled by altering concentration of tetracycline in the drinking water of mice.

The most important feature of a DNA construct is that the DNA must integrate into the host chromosomes to give stability and heritability. If the DNA remains in the cytoplasm then it will be expressed only transiently and lost during cell division. The DNA forms concatomers that will integrate into one site in the genome. However, the site of integration is random and this could be a major problem, for example if the construct is integrated into a gene for the synthesis of a vital enzyme. Such integration will de-activate the host gene; this is termed 'insertional mutagenesis'. This is undesirable and may kill the transgenic animal. It is possible to direct the construct to a specific site in the genome using a method called 'homologous recombination'. This relies on replacing a piece of host DNA with the construct; it is difficult but has been used successfully in the construction of mouse models for sickle cell disease with the normal β-globin gene being replaced by the mutated form.

The principle underlying homologous recombination is the reliance on the recombination that may occur between the host chromosomal gene and a homologue – in this case our construct. The construct may be taken up and exchanged for the host gene. This method works when the gene construct is very similar to the host genes, although we need to test all the transgenics to determine which, if any, has the construct in the appropriate site. This is more difficult if we wish to incorporate a novel sequence into the host genome at a specific site. In order to do this we must then engineer the construct so that at each end it contains sequences found in the host at the site of the required insertion. Success is poor with this method.

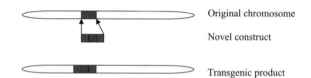

In this case the construct has enough homology with the gene on the host chromosome to be taken up in a crossing-over event.

Figure 7.2   Homologous recombination

7.5 Introduction of the DNA into the cells

Many methods have been described for introduction of the DNA into cells, but most are not suitable for the introduction of DNA into single egg cells; that is, for germline gene therapy. We need a highly efficient and safe methodology since we have limited numbers of eggs and we need to retain viability. Calcium phosphate and electroporation, methods which are often used to transfect cultured cells, are of little use because of low efficiency. The most common method of DNA introduction into eggs is the micro-injection technique. Although a tedious technique it is ideal when we have only a few eggs to manipulate. It is highly efficient when we are working with rodents since the male pronucleus is obvious. However, with larger mammals the technique is less reliable as the cytoplasm of the egg is opaque and the male pronucleus is difficult to identify. Researchers have investigated the use of dyes to highlight the pronucleus but these cause problems with cell viability. Therefore, viral vectors may hold more promise with larger mammals. Liposome delivery has been attempted but has no real advantages in this case. Viral vector systems are also used for delivering DNA in the embryonic stem cell method.

7.6 Uses of transgenics

7.6.1 Recombinant protein production

Using transgenic animals to produce recombinant proteins has many advantages, such as the fact that we can target expression of the desired product encoded by the transgene to the animals' milk. This makes the protein easy to collect and purify and enables us to produce large quantities of the recombinant protein. This also means that once a producing female is found we can breed more transgenic females to produce a herd of transgenics where all females produce the recombinant protein.

The Roslin Institute and PPL Therapeutics have already used the embryonic stem cell approach to produce transgenic animals more efficiently than is possible with microinjection. PPL has incorporated the gene for human Factor IX, a blood-clotting protein used to treat hemophilia B into sheep. They transferred an *antibiotic-resistance* gene to the donor cells along with the *Factor IX* gene, so that by adding a toxic dose of the antibiotic neomycin to the culture, they could kill cells that had failed to take up the added DNA. In 1997 Polly[115] was the first transgenic sheep produced in this way. Polly and other transgenic clones secreted the human Factor IX protein in their milk.

Once techniques for the retrieval of egg cells in different species have been perfected, cloning (see Chapter 8) will make it possible to introduce precise genetic changes into any mammal and to create multiple individuals bearing the alteration. The use of transgenic animals, particularly larger mammals, as bioreactors, 'pharmaceutical pharming', is a cost-effective alternative to cell culture methods. Animals automatically supplement their bodily fluids with fresh nutrients, remove waste products, reliably regulate their internal temperature and pH and resist pathogens. By targeting the expression of the transgene product so produced by the secretory cells of the liver, lactating mammary gland or kidney, 'pharmers' may collect and process bodily fluids with minimal effort. The mammary gland is the most promising target tissue as it produces large amounts of protein that can be collected daily in a non-invasive fashion.[116]

Transgenic animals can perform much more complicated protein modifications than cultured cells although there can still be problems with the correct processing of the protein product, for example some complex post-translational modifications of proteins are seen in humans and the higher apes only. Transgenic sheep have been developed which make recombinant protein in their milk in larger quantities than could be produced by conventional cell-culture methods. However, there is evidence that biological products can leak into the blood supply of the animal that is producing them. Although the leakage of certain products, for example α-1-antitrypsin, does not represent a risk to the animal, other products could interfere with normal bodily functions.

The main disadvantages associated with using transgenic animals to produce recombinant proteins are the initial cost and time involved in establishing a successful transgenic animal that produces the desired protein. Although once a successful animal has been produced the yields are much better compared to recombinant proteins produced through cell culture, for example the production of anti-thrombin in transgenic goat's milk yields 7 g per litre of milk compared to 50 mg per litre in tissue culture. The overall costs are 50% of those of recombinant proteins made in cell culture and the proteins are easier to purify, but quality assurance (QA) and quality control (QC) are a problem as there is no consistency in the product. Examples of recombinant, made in transgenic animals are α-1-fetoprotein made in goats, albumin and anti-thrombin made using YAC as vectors in goat ES transduction[117] and Factor IX produced in pigs.[118]

7.6.2 Animal models of human disease

Animal models of human disease are made by manipulating the animal's genome in order to produce an animal that mimics a human disease. Animal models are important in the study of disease mechanisms and the trailling of new therapies and treatments and often form a crucial part of late-stage pre-clinical studies. Most genetic diseases can be categorised in terms of their molecular basis. In loss-of-function mutations, protein is not produced, resulting in disease. In gain-of-function mutations, a mutation alters the resultant protein from its native one, causing disease. Recessive inherited traits tend to be loss-of-function mutations and dominant-inherited traits tend to be gain-of-function traits. If we want to model a recessive trait we will need to mimic a loss-of-function mutation. In this case, when making the transgenic animal, the aim is to make a knock-out animal where we produce a null allele for the gene of interest in our construct and insert this into the mouse gene equivalent using the ES cell method, relying on homologous recombination for correct localisation of the gene to replace the functional copy with the null allele. Following injection of ES cells into the blastocyst, the transgenic mouse would be allowed to develop and breed, and then the offspring would be screened. The transgenic animal model for beta thalassaemia was produced in this way.

If we want to model a dominant trait then we are most likely to model a gain-of-function mutation. In this instance the construct will contain a mutant gene whose presence is sufficient to induce the desired disease. The most suitable method is pronuclear microinjection and the disease to be modelled must be one where the presence of introduced DNA is sufficient to produce pathology. It would be necessary to first isolate the mutant gene or design it by *in vitro* mutagenesis. The construct containing the mutant gene would be transferred to the fertilised oocyte, unlike when we are making the loss-of-function model; in the present case there is no requirement for a specific location of integration of the construct. The model for spinocerebellar ataxia, a neurodegenerative disorder that arises from triplet expansion (CAG) in the *ataxin* gene (*SCA1*), was made in this way.[119] Transgenic mice were produced with the normal human *ataxin* gene and one with the expanded repeat; both transgenics were stable, but only the transgenic mouse produced with the construct that had the expanded allele developed ataxia and Purkinje cell neurodegeneration.

One of the major disadvantages of using transgenic animals as models of human disease is that sometimes the symptoms of the disease observed in the animal model are not the same as those seen in humans. Additionally, the lifespan of many animals is shorter than that of humans, their genomes are different and so interactions may occur between genes in our animal models that will never be seen in humans. Often results obtained from animal models are complex and not all therapies that work in mice work in humans. Some drugs, for example thalidomide or the antibody used in the recent Parexel trials[120] in the UK, were harmless to mice but devastating to humans. We need to be careful, as often we do not understand the relevance of the results we obtain.

There are a number of databases that house the complete list of transgenic knock-in and knock-out mice available. Many list references and both genotype and phenotype data.[121]

Points to consider

There are questions raised about the usefulness of transgenic animal models in disease therapy. There are many who do not think creation of transgenic animals is right. Indeed, we have banned germline therapy, the production of transgenics, in humans. Some things we need to consider include: the benefits, safety concerns, environmental impact, 'unnaturalness': the devaluation and commercialisation of animal life, animal welfare concerns,[122,123] inefficiency of the technique, mutation induced, inappropriate gene expression and cost effectiveness. Clearly, transgenics have a role in modelling of diseases and trialling of therapies, but as yet they have not fulfilled their projected usefulness in production of recombinant proteins.

Points to consider

Do the general public approve of animal models? Ask your friends and families.
Do you get different answers dependent on the type of animal used?
View the video available at http://ihome.cuhk.edu.hk/%7Ez045478/bioch

Notes

111 Higgins, C.F. and Trezise, A.E.O. Cystic fibrosis mice have arrived. *Human Mol. Genetics*, 1992, **1**, 459–60
112 Koehler, D.R. Protection of *Cftr* knockout mice from acute lung infection by a helper-dependent adenoviral vector expressing *Cftr* in airway epithelia. *PNAS*, 2003, **100**, 15364–9
113 http://www.drhadwentrust.org/
114 Labosky, P.A. et al. Mouse embryonic germ like growth factor 2 receptor (Igf2r) gene compared with embryonic stem (ES) cell lines. (EG) cell lines: transmission through the germline and differences in the methylation imprint of insulin-like growth factor 2 receptor (Igf2r) gene compared with embryonic stem (ES) cell lines. Development, 1994, 120, 3197–204
115 Colman, A. Dolly, Polly and other 'ollys': likely impact of cloning technology on biomedical uses of livestock. *Genet. Anal.*, 1999, **15**, 167–73
116 Baldassarre, H. *et al.* State of the art in the production of transgenic goats. *Reprod. Fertil. Dev.*, 2004, **16**, 465–70
117 Zhang, XF. *et al.* Transfer of an expression YAC into goat fetal fibroblasts by cell fusion for mammary gland bioreactor. *Biochem. Biophys. Res. Commun.*, 2005, **333**, 58–63
118 Lindsay, M. *et al.* Purification of recombinant DNA-derived factor IX produced in transgenic pig milk and fractionation of active and inactive subpopulations. *J. Chromatogr. A.*, 2004, **1026**, 149–57

119 Burright, E.N. *et al.* SCA1 transgenic mice: a model for neurodegeneration caused by an expanded CAG trinucleotide repeat. *Cell*, 1995, **82**, 937–48
120 Editorial. What really happened in drug trial disaster? *New Scientist*, 19 August 2006, 9
121 www.deltagen.com/
122 www.bbsrc.ac.uk/tools/download/ethics_animal_biotech/ethics_animal_biotech.pdf
123 Dunn, D.A. *et al.* Foundation review: Transgenic animals and their impact on the drug discovery industry. *Drug Discov. Today*, 2005, **10**, 757–67

8

Transplantation: a form of gene therapy

Terminology

An *autograft* is a transplant from the patient concerned; this is typically bone marrow, bone or skin. An *allograft* is a transplant from someone else; this can be a living donor (ALD), cadaver donor (AC) or from an aborted fetus (AF). A *xenograft* is a transplant from a different species.

8.1 Introduction

It can be argued that transplantation is a form of gene therapy; the basis of gene therapy is the delivery of functional genes to the patient, and in transplantation of allografts, the patient receives new cells with new genetic information. The genetic information introduced by the transplant has all its own natural control mechanisms and is delivered with an undamaged fully functioning organ. Common transplants include bone marrow, blood and organs including the kidney, heart, lung, liver and pancreas.

Molecular Therapeutics: 21st-century Medicine by Pamela Greenwell and Michelle McCulley.
© 2007 John Wiley & Sons, Ltd

8.2 Bone marrow

Bone marrow contains stem cells that produce the different types of blood cells: white blood cells that fight infection, red blood cells that carry oxygen to and remove waste products from organs and tissues, and platelets that enable the blood to clot. Bone marrow transplants (BMT) are probably the most common form of transplantation. In 1980 bone marrow transplants were carried out in just two or three specialist centres in the UK; today more than 300 allografts and 600 autografts are performed annually in more than 50 hospitals.

BMT is offered to many leukaemic patients who have entered remission from their disease, replacing their potentially cancerous marrow with marrow from an unaffected individual (allograft) or more commonly bone marrow may be removed from patients in remission, for example leukaemia and lymphoma sufferers, treated with antibodies to remove any cancerous cells and then frozen and returned, in case of relapse (autograft). Bone marrow support transplants may also be needed in cases of patients receiving toxic treatment for cancers such as breast cancer. In this case, bone marrow function is repressed by the therapy and an autograft could be vital. Hereditary defects of the immune system such as such as severe combined immune deficiency (SCID) or adenosine deaminase deficiency (ADA) may be cured with allogenic transplants. Indeed, one criterion for acceptance onto the gene therapy trials for this disease is that there is no known marrow donor available. Aplastic anaemia may be treated with BMT as well as a series of hereditary diseases such as thalassaemia, sickle cell disease, and Gaucher's and Batten's diseases.[124] In these cases BMT is not considered unless a child is not responding to any of the current therapies available, since BMT is itself a risky procedure.

The success of allo-transplantation depends on human lymphocyte antigens (HLAs)[125] matching; only 30 to 40 per cent of patients have an HLA-matched sibling or parent. The chances of obtaining HLA-matched marrow from an unrelated donor are small with allo-BMT; a complication known as 'graft-versus-host disease' (GVHD) sometimes develops. GVHD occurs when white blood cells from the donor marrow (the graft) identify the cells of the patient's body (the host) as foreign and attack them. GVHD can generally be treated with steroids or other immunosuppressive agents.

Typically, bone marrow is collected from the pelvic bones whilst the donor is under general anaesthetic; usually, a small cut is made in the skin over the pelvic (hip) bone, a large needle is inserted through the cut and into the bone marrow to draw the marrow out of the bone. The harvested bone marrow is then processed to remove blood and bone fragments. Most stem cells are found in the bone marrow, but stem cells called peripheral blood stem cells (PBSCs) can be found in the blood stream at much lower concentration. Recently, peripheral blood has largely replaced marrow as a source of stem cells for autografting. Stem cells found in peripheral blood can be harvested after treatment of donors with mobilising agents, such as colony-stimulating factors (CSFs), which encourage stem cells to enter the peripheral blood; in this case blood stem cells are extracted by apheresis and the donor does not need to be under general anaesthetic. There have, however, been some concerns, following haemopoeitic changes seen in a small number of donors, as to the safety of CSF administration in normal donors.

Prior to receiving the marrow, the patient's own marrow is destroyed by radiation and chemotherapy. Blood components, such as platelets, are also given to the patient after transplant in order to maintain health. After entering the blood stream, the transplanted cells travel to the bone marrow, where they begin to produce new white blood cells, red blood cells and platelets in a process known as 'engraftment'. This usually occurs within about 2 to 4 weeks after transplantation. Complete recovery of the immune function takes much longer, up to several months for autologous transplant recipients and 1 to 2 years for patients receiving allogeneic transplants. Biotechnology companies are working to grow and stimulate stem cells in culture for transplantation. In the hospital environment stem cell therapy cuts hospital in-time when compared to a traditional bone marrow transplant.[126,127]

Umbilical cord blood also contains stem cells and could be used instead of adult bone marrow, although it is usually used to treat under-16.[128] Cord blood has advantages in that it has little in the way of antigens expressed on the blood cells and rejection such as GVHD appears to be far less of a problem. The first cord blood transplant took place in France in 1988 for a child with Fanconi's anaemia. Collection is non-invasive, painless, harmless and quick, and in theory approximately 50 ml can be harvested at each delivery for use in transplantation. NHS facilities within the National Blood Service have been undertaking cord blood banking since 1996; the cord blood is donated to the banks in a similar way to bone marrow from donors and hence use is limited by the need to find an HLA-matched donor.

Cord blood banking in the UK is undertaken only at the following hospitals: Northwick Park Hospital in Harrow, Barnet General Hospital, Newcastle Royal Infirmary and the Mater Infirmorum Hospital in Belfast. Some transplant centres currently recommend cord blood collection and storage for siblings born into a family where there is a known genetic disease amenable to HSC (haemopoietic stem cell) transplantation. The cord blood banks are different to autologous cord blood storage now being offered commercially in the UK where cord blood can be stored as the property of the child for use later in life (private companies such as UK Cord Blood Bank and more recently Virgin offer this service). Another relevant issue is the question of designer siblings much documented in the media.[129,130] The procedures involved IVF and PIGD to produce siblings without disease, but good HLA matches for already affected children. The press has dubbed these siblings as 'designer babies', though, as discussed previously, we should more correctly view them as 'selected babies'.

BMT has also been used in an attempt to treat HIV, with variable results. There have been suggestions that genetically modified cells could be used to fight the infection.[131]

8.2.1 Logistical problems with BMT

The three main problems with bone marrow transplants are supply, matching marrow and lack of donors from ethnic minorities. Every year in the UK more than 9000 patients are diagnosed with diseases that can be treated with BMT but ~70% are unable to find a donor. The problem is more apparent in ethnic minorities, for example the need for organs in the Asian community is 3–4 times higher than in Caucasians. Bone marrow registers such as the Anthony Nolan Trust (www.anthonynolan.org.uk) in the UK encourage registration and have a database of donors for transplantation.

8.3 Solid organ transplantation

Professor Sir Magdi Yacoub has described transplantation as 'one of the great success stories of the latter half of the 20th century'. Key dates in transplant history include the first transplant, which was a cornea in 1905, the first transplanted organ, which was a kidney in 1954, and the first heart transplant carried out in 1964 by Christiaan Barnard in South Africa. The UK's NHS Organ Donor Register was launched in October 1994 and by November 2002 included the names of more than 10 million people. A single individual who donates after death can provide organs, corneas, skin, bone and tissue for 80 people in need. As a minimum for solid organ transplantation for the allogeneic or unrelated stem cell recipients, the patient and short-listed potential donors must be typed for *HLA-A, -B, -C, -DR* and *-DQ* gene products by molecular methods at high resolution to facilitate the final selection process. Preventing disease is one way to deal with the demand for organs, tissues and cells for transplant. Exercise, healthy eating and reduction in the use of alcohol and tobacco are all ways to reduce disease and the need for a significant proportion of organ transplants.

8.3.1 Heart transplantation

The first heart transplants were carried out in the 1960s, but few of these were successful. In part this was because patients selected for this risky and experimental therapy were already extremely ill and therefore poor candidates, and anti-rejection drugs were still in their infancy.

Table 8.1 Heart transplant history

- 1st heart transplants in dogs in the 1950s with a maximum survival of 165 days in dogs treated with immunosuppression and 7 days for those not receiving drug therapy. Mean survival = 103 days
- 1st human heart transplant in 1967 died 18 days later.
- 2nd transplant patient lived 19 months and the third 20 months
- between December 1967 and June 1968, 21 transplants were carried out with a 1-year survival of 22%
- in 1968, 105 transplants took place and 65% of patients who were treated before 1970 died within three months
- in 1976 the 1-year survival rate had increased to 67%; it is currently 85%. Maximum survival recorded = 20 years
- 10-year survival is now 45%

Reliable cardiac transplantation began in the UK in 1979. Combining the anti-rejection drug cyclosporine with steroids and other drugs produced excellent results. Heart transplants are used to treat those with congenital heart deformities, heart failure and virally damaged hearts. Heart transplants are almost always from young brain-dead accident victims. Interestingly, seat-belt laws have decreased the number of heart donors. Usually heart transplants are allografts from cadavers (AC) but they are occasionally from living donors (ALD); typically an ALD would be a CF sufferer who has severely damaged lungs and is receiving a heart–lung transplant and so their own heart can be transplanted to another patient. Life expectancy for heart transplantation can be 20 years and about 60% of patients live for 5 years post-transplant.

There is a severe shortage of donors, and statistically patients are more likely to die whilst waiting for an organ than in the first year post-transplant. Possible post-operative complications may arise following heart transplant surgery. They include vascular problems (bleeding); arrhythmias (irregular heart rhythm); lung problems (collapsed lung; pneumonia); and incision problems. Other problems include the long-term risks of immunosuppression, and increased risk of cancer and infection; this is common to all transplantation and is a direct result of the immunosuppressive regime.

8.3.2 Lung

The first single-lung transplant was performed in UK in 1987 (bilateral lung in 1990), while the first combined heart–lung was performed in 1984. Difficulties in the combined transplant are very rare. Lung transplantation has been used to treat end-stage lung disease in inherited and acquired disorders such as CF.[132] Patients with CF usually have both heart and lungs replaced in the surgery, freeing their healthy heart for subsequent donation. Although most of the transplants involve organs from cadavers, some are from live donors.[133] Parents of children with CF may donate part of their lung for partial transplant. Indeed, in some cases both parents have provided lung sacs for grafting. In one of the first such operations carried out in the UK the young girl concerned was cured of her breathing problems but during the operation suffered a stroke that left her paralysed. There are significant worries about the ethics and morals in this type of case where the parents feel guilty, both risk their lives and the child is no better.

8.3.3 Kidney

These may be taken from cadavers or live, usually related, donors.[134] It was the first successful solid organ transplant, and it is now one of the safest types of transplant and the most cost-effective form of treatment for patients in end-stage renal failure (ESRF) seen in diabetes and polycystic kidney disease. Pre-treatment of the organ prior to transplant, for example with monoclonal antibodies (see Chapter 6), has in turn increased the success of the technique. Unfortunately, the supply of donor organs, which averages only 30 per million patients (pmp) per year in UK, is greatly outstripped by the demand (48 pmp). Transplants are effective for decades; a kidney transplant can last from 8 to 25 years. Living donor results are even more successful: studies show that fewer than 4 deaths occur out of 10,000 donors. Kidney transplant recipients usually need to take immunosuppressants, or anti-rejection drugs, for the rest of their lives, and benign skin cancers that are frequently seen in these patients are a direct result of the immunosupression which allows human papillomavirus, which is normally not a problem, to proliferate and result in cancerous growths. Recently the ethical issues surrounding live donors have been emphasised by 'organs for sale' reported in the media involving donors from poor countries earning a lifetime's wage by selling a kidney. Another individual has reputedly earned millions of dollars through selling a kidney on an internet auction. Thus, in the UK, there has been a change in the law enabling non-related donations provided there is no financial gain.

8.3.4 Liver

Liver transplants can either be live partial (ALD) or from a cadaver donor (AC). The liver is unique in that when damaged it can regenerate: live donors can donate portions of their liver or a donated liver can be split to be used in a number of patients. Conditions referred for transplantation include primary biliary cirrhosis, chronic active hepatitis, alcoholic liver disease, cirrhosis resulting from hepatitis B and C, cancer and fulminant hepatic failure (including paracetamol overdose). The issue of treating patients with hepatitis is interesting since, if the virus is still present in the patient, the new liver may also be destroyed. Liver transplantation is not usually performed when people have liver cancer, if cancer has already spread.

8.3.5 Ovarian tissue

In 2004 the first baby was born after a woman had undergone an ovarian transplant. The woman had chemotherapy for hodgkin's lymphoma in 1997 when she was aged 25, doctors took five small strips of tissue from her left ovary before she started the cancer treatment and froze them. Tests confirmed her ovaries had stopped functioning. Doctors re-implanted a strip and smaller cubes of ovarian tissue back into the patient's ovaries. Five months later, the patient started to have normal periods again and became pregnant. There are a few ethical issues surrounding this case, however ovarian tissue could also be harvested from aborted fetuses and used to treat women who are infertile. This is banned in the UK because of the perceived ethical issues associated with the birth of children whose mothers were in fact aborted fetuses.[135]

8.4 Other cells and tissues

Bone has been harvested from cadavers for many years for use in reconstruction surgery to treat patients with inherited diseases such as Crouzon syndrome. In bone injury, autografts may also be used.[136] Skin is required by burns victims, accident victims and in reconstruction surgery for inherited deformities. In the past skin was removed from an unaffected part of the body, donated by live skin donors or taken from cadavers. The main problem with skin transplants is that the skin resembles the area from where it was taken. It is possible now to grow skin as a monolayer in culture that can then be used in surgery. Reports suggest that $1\,cm^2$ of foreskin tissue removed at circumcision could be grown in culture to produce enough skin to cover a football pitch! The skin takes on the characteristics of normal skin from the area being treated. This reduces scarring and makes the grafts far less obvious.

Corneal transplantation restores sight to recipients and is one of the most successful transplantations. Approximately 3000 corneal grafts are carried out each year in the UK. Corneas can be removed after death, unlike other organs that need to be removed whilst the patient is kept alive artificially by machines; therefore people are often more willing to donate corneas. Nevertheless there is still a shortage of donors and researchers have used biotechnology to make up the shortfall. It is now possible to take a small area of corneal tissue, grow this in culture and then use that material for engraftment. Eye banks retain corneas for about 10 days at 4 °C and use material from organ culture for up to 30 days. There have been reports of transmission of CJD through corneal grafts. Donors are not routinely tested for CJD as the test can involve SDS-PAGE and Western blot which is time-consuming.[137]

Pancreatic transplants could provide a cure for insulin-dependent diabetes. Few, however, are carried out since diabetes may be controlled by recombinant protein therapy (with insulin).[138]

Research has been carried out where brain cells harvested from fetuses at abortion have been implanted into the brains of individuals suffering from diseases such as Parkinson's and Huntington's.[139] In the first double-blind, placebo-controlled surgical trial testing the safety and effectiveness of fetal dopamine cell implantation for the treatment of Parkinson's disease, most patients who received the implants showed growth of the new brain cells, and many had improvement in their symptoms, though some had long-term complications.[140] Fears have been raised that some women could produce fetuses for abortion and transplantation in exchange for large sums of money.

8.5 Summary of the problems associated with transplantation

Donor shortage remains the greatest problem; there is some debate whether this could be improved by exchanging the system of having voluntary organ donation cards for an opt-out system. Currently in the UK, we have a card system, but many do not realise that the next-of-kin's permission must be obtained before organs can be removed. A dilemma in an over-stretched hospital is whether a patient who is not likely to recover should be maintained on life support simply to enable doctors to locate their next-of-kin to gain permission for organ harvesting. In other countries there is an opt-out system whereby everyone is seen as a potential donor unless they have specifically opted out. This system is currently being reviewed and may yet be adopted in the UK.[141]

Fetal donation could be used for some transplants; however, this is a contentious issue and remains banned in many countries. Similarly ethical issues shroud the concept of live donors, both for safety reasons and also to ensure that emotional (family) and financial payment pressures are avoided.

Absolute contraindications to organ donation are HIV 1 and 2; human T-cell lymphotropic virus (HTLV) 1 and 2; systemic viral infections such as hepatitis B and C; deep or systemic fungal infections; prion disease and resistant organisms (e.g. MRSA, VRE – vancomycin-resistant enterococci). However, the health of donors prior to death can be uncertain: there are reports of HIV, hepatitis B, CJD and cancer being transmitted with organ transplant. On 16 February 2005, the Deutsche Stiftung Organtransplantation (German Foundation for Organ Transplantation)[142] reported possible rabies in three of six patients who received organs from a donor who died in late December 2004. It is possible to test for HIV antibodies within about 2 hours but this will not detect early infection. PCR-based testing may take 5–6 hours and, undoubtedly, this contributes further anxiety to the already stressed relatives of donors. In the US, there are reports of patients suing next-of-kin of donors for allowing infected material to be transplanted. Although relatives may be questioned as to the risk factors associated with disease transmission, many are not fully aware of the activities or personal lives of their relatives.

8.6 Transplantation statistics

The UK transplant website contains the most up-to date information on statistics available for the UK.[143]

In the USA 45,000 patients would benefit each year from heart transplants but less than 2000 hearts are available; 24,000 patients are awaiting a kidney transplant. In the UK 6000 patients are waiting for transplants, 5000 for kidneys, 1000 for hearts, lungs or livers. The list grows at 5% per year and only half receive the organ they need.

8.7 Legislation[144,145]

Legislation exists to prevent the misappropriation and misuse of organs and to allow transparency in the system. It should prevent misuse of organs and coercion of donors. Allocation of organs must be organised in an equitable way. Currently some organs are more readily available to private patients than those seeking treatment through the NHS in the UK. Some groups are denied organs, for example Down's children (due to associated problems of predisposition to malignancy), heavy smokers and alcoholics. Yet liver transplants can be allowed on patients overdosing on paracetamol or dying from alcohol-induced liver failure. In the UK, there was indignation when the footballer George Best, following a liver transplant for alcohol-induced liver failure, resumed his consumption of alcohol and died. A similar situation was reported with the actor Larry Hagman. It was reported in February 2002 that a prison inmate in California received a heart transplant, and this sparked debate across the US. Japanese law concerning organ procurement is the most stringent in the world. It does not allow doctors to take organs from a brain-dead person unless he or she left written consent and the family agrees with the donation. Moreover, the law stipulates that only people aged 15 and older can give consent, presuming it too difficult for children to make their own decisions. As a result, the likelihood of children receiving hearts of the proper size is next to nil.

In the UK, much concern has been expressed about tissue donation and storage. This follows a number of high-profile cases in which organs from children were removed and kept without parental consent.[146] The Tissue Act 2004 has been superseded by the legislature of the Human Tissue Authority.[147] The Human Tissue Act 2004 (HT Act) came into force on 1 September 2006 in England, Wales and Northern Ireland. The HT Act regulates the donation by living people of solid organs, bone marrow and stem cells and of solid organs from cadavers. There is also a separate act, the HT (Scotland) Act, that is almost identical to the HT Act. That takes responsibility for approving all forms of living donation involving Scotland. The HTA is responsible for approving all living-donor transplants. The Act also lays down guidelines for the use of materials where the deceased has made their wishes clear, but their family disagrees: essentially they will act on the wishes of the deceased.

Recent reports have suggested a change in the legislation concerning the definition of death with some hospitals reportedly using cessation of cardiac function (DCD – donation at cardiac death), that is when the heart ceases to beat but there is still potentially brain activity, rather than brain-stem death as an indicator. This is prompting debate as to the nature of death and whether, given the albeit low numbers of patients recovering from persistent vegetative state or coma, cessation of cardiac function, in the absence of life support intervention, is a valid definition. It is suggested that changes to the law to allow DCD donors would increase the numbers of organs available by 20%.

Table 8.2   The pros and cons of transplantation

ADVANTAGES
• Transplantation offers a CURE.
• It is gene therapy in disguise; we are replacing bad genes with good genes when we carry out the organ replacement.
• In transplantation, genes are added with all their control mechanisms.
• The new organ bears no scars from the patient's disease.

DISADVANTAGES
• Supply.
• Each organ must be matched to the patient's – some patients have such rare antigens that they are almost impossible to match.
• Rejection, i.e. graft versus host and host versus graft can lead to organ failure and death.
• Patient treated long term with immunosuppressants, e.g. cyclosporine, leading to risk of cancer.
• Surgical procedure: skilled, expensive and dangerous.
• Continuous monitoring and drug regime required.
• Risk of infection from organ, e.g. HIV, hepatitis.
• Risk of CJD.
• Risk of cancer from organ.
• Cost.

8.8 Religious beliefs and transplantation[148]

All major religions support organ and tissue donation. Though beliefs and doctrines vary slightly among various denominations, the underlying theme is the same: organ and tissue donation represents one of the highest forms of loving, giving and caring – the principles upon which all religions are based. The Protestant faith respects an individual conscience and a person's right to make decisions regarding his or her own body. Catholics view donation as an act of charity, fraternal love and self-sacrifice. Transplants are ethically and morally acceptable to the Vatican. The Greek Orthodox Church is not opposed to organ donation as long as the organs and tissue in question are used to better human life, i.e. for transplantation or for research that will lead to improvements in the treatment and prevention of disease. Islam strongly believes in the principle of saving human lives. According to A. Sachedina in *Transplantation Proceedings*, 'the majority of the Muslim scholars belonging to various schools of Islamic law have invoked the principle of priority of saving human life and have permitted the organ transplant as a necessity to procure that noble end'.

All four branches of Judaism (Orthodox, Conservative, Reform and Recon-structionist) support and encourage donation. Organ donation is the only *mitzot*, or good deed, an individual can perform after death. Hindus are not prohibited by religious law from donating their organs. This act is an individual's decision. Buddhists believe that organ or tissue donation is a matter of individual conscience and place high value on acts of compassion. In Shinto, the dead body is considered to be impure and dangerous, and thus quite powerful. It is difficult to obtain consent from bereaved families for organ donation or dissection for medical education or patho-logical anatomy. To Mormons, the body is seen as essential, 'part of the soul', but there is no prohibition against donating or receiving organs in appropriate medical circumstances. Procurement from a live fetus that would prevent it from live birth would be opposed.[149] Jehovah's witnesses believe that there is no biblical command to prevent organs being donated or received, but to do so is a personal decision.[150]

Notes

124 http://adc.bmj.com/cgi/reprint/90/12/1259.pdf
125 www.anthonynolan.org.uk/index.php?location=4.3
126 http://linkinghub.elsevier.com/retrieve/pii/S0188440903001000
127 www.jcmm.org/en/pdf/9/1/jcmm009.001.05.pdf
128 http://linkinghub.elsevier.com/retrieve/pii/S0268960X0300064X
129 Pennings, G. *et al.* Ethical considerations on preimplantation genetic diag-nosis for HLA typing to match a future child as a donor of haematopoietic stem cells to a sibling. *Human Reproduction*, 2002, **17**, 534–8
130 www.timesonline.co.uk/tol/news/uk/article501875.ece
131 www.medscape.com/viewarticle/413256?src=search
132 http://www.cff.org/treatments/LungTransplantation/
133 www.medscape.com/viewarticle/458740_print
134 http://www.uktransplant.org.uk/ukt/statistics/latest_statistics/latest_statistics.jsp
135 www.ivanhoe.com/channels/p_channelstory.cfm?storyid=11568
136 www.blood.co.uk/foi/13_Policies_and_procedures/Long_term_strategy_for_provision_of_tissue_banking_facilitiesNOV02.pdf
137 http://bmj.bmjjournals.com/archive/7122/7122e1.htm
138 http://jcem.endojournals.org/cgi/content/full/82/8/2471
139 www.hda.org.uk/research/rs025.html
140 Freed, C.R. *et al.* Transplantation of embryonic dopamine neurons for severe Parkinson's disease. *NEJM*, 2001, **344**, 710–19
141 http://news.bbc.co.uk/1/low/health/384401.stm

142   www.dso.de/

143   www.uktransplant.org.uk/ukt/statistics/transplant_activity_report/current_
      activity_reports. jsp/ukt/tx_activity_report_2005_uk_complete-v2.pdf

144   www.uktransplant.org.uk/ukt/about_transplants/legislation/legislation.jsp

145   www.parliament.uk/documents/upload/POSTpn231.pdf

146   www.rlcinquiry.org.uk/download/chap10.pdf

147   www.hta.gov.uk/

148   www.xeno.cpha.ca/english/viewpnt/relig/relig.htm

149   www.jmahoney.com/mormon.htm

150   www.kyha.com/documents/CG-JW-REV.pdf

9

Xenotransplantation

9.1 Introduction

Xenotransplantation is any procedure that involves the transplantation, implantation or infusion into a human recipient of live cells, tissues or organs from a non-human animal source, or human body fluids, cells, tissues or organs that have had *ex vivo* contact with live non-human animal cells, tissues or organs. Ideally this should take place between closely related species to avoid severe rejection problems.

Xenotransplantation has a longer history than human-to-human transplantation. The transfer of blood from animals to humans was popular prior to the discovery of blood types and safe transfusion methods. The first serious animal-to-human transplants, from chimpanzees, pigs and sheep, took place in the 1960s prior to the first successful human–human transplants. These all resulted in the patient's death. In 1984, Baby Fae was given a baboon heart and lived for 20 days. Prior to this, a number of baboon-to-goat and lamb-to-goat transplants had taken place, with a mean survival time of 72 days. In 1992, a pig heart was transplanted into a patient with Marfan's syndrome – death occurred after 24 hours. In the same year Thomas E. Starzl, one of the pioneers in transgenic transplants, transferred a baboon liver into an HIV patient suffering from hepatitis B. The patient survived for an agonising 70 days. It was reported that Starzl considered the transplantation to have been a success, though the patient's death makes gruesome reading:

Molecular Therapeutics: 21st-century Medicine by Pamela Greenwell and Michelle McCulley.
© 2007 John Wiley & Sons, Ltd

> *By turns, he suffered from septic intoxication, oesophagitis, viraemia (the presence of viruses in the blood), haemorrhaging in the pleural (chest) cavity, and later from circulatory collapse, as well as an acute cough. In the end, kidneys and liver failed, and bile engorgement was produced. The patient finally died from internal bleeding.*[151]

9.2 Rationale for the use of non-human donors

Using animals as organ, tissue or cell donors could circumvent many of the problems associated with organ donation, which could mean that there would be no shortage of donor organs and therefore even high-risk groups could be treated. Additionally, animals are often immune to diseases seen in humans, for example hepatitis. The following section discusses candidate animals for xenotransplantation, problems with xenotransplantation and future applications.

9.3 Organs from non-human primates

The nearest relatives to humans are the bonobo chimpanzees. However, their use in xenotransplantation is an emotive issue. They are also endangered in the wild and relatively few are kept in captivity. The cost of raising them could be very high as they breed slowly, take many years to reach sexual maturity and are not fully grown for 5–6 years. Bonobos are also small and their organs would only be suitable for children in the short term and would presumably need to be replaced with human organs as the child grew. There are over 20 known potentially lethal viruses that can be transmitted from non-human primates to humans, including Ebola, Marburg, hepatitis A and B, herpes B, SV40 and SIV. They also may transmit to humans diseases such as green monkey disease, tuberculosis, encephalitis, meningitis and haemorrhagic fever. This may lead to the spread of disease from the recipient into the general population and would result in a public health problem. Virus-free breeding stocks would be difficult to set up and maintain. Primates also have different percentages of ABO blood types from humans, for example no baboons have been described with blood group O: the most common human blood type. This would mean that only blood types A, B and AB could be treated with baboon organs.[152]

9.4 Pigs

The pig reproduces quickly and as it is an animal that is already bred for food there is likely to be less of a problem in terms of animal rights for their use in xenotransplants.[153] Pigs have organs of similar size to humans, for example pig hearts are a suitable size for humans whereas sheep and goat organs are too small and those of cows too big. Pigs have a relatively long lifespan and blood groups A and O occur in pigs. Since O is a universal blood group donor there should be no rejection problem based on ABO blood groups. Pig valves have also been used for decades to replace non-functioning human heart valves.

Nevertheless, pigs are very different from humans. If we transplant a pig organ into a human we will see an almost instant response: the organ will darken and cease functioning. This is known as 'hyper-acute rejection' (HAR). HAR occurs due to the activation of the complement system by the attachment of antibodies to the endothelial cells of the xenotransplant. This triggers the complement cascade, resulting in intravascular coagulation, oedema and catastrophic destruction of the organ. Humans have circulating xenoreactive natural antibodies (XNAs), so called because they recognise the cells of foreign species and are present despite a lack of previous exposure. More than 85% of human XNAs are targeted to a specific sugar residue, Galα1-3Gal, which is present on the endothelial cells. The binding of XNAs to the Galα1-3Gal residue leads to the activation of the classical complement pathway. The complement cascade is known to be regulated by species-specific regulators (RCA – regulators of complement activation). Thus, if pigs could be engineered, using transgenic technology to contain the human version of one RCA, DAF (decay-accelerating factor), the problem could be solved. Genetically engineered pig organs have been tested only in baboons in which non-transgenic hearts were destroyed in 55 minutes whereas transgenic hearts were rejected after a mean of 90 days.[154]

Numerous strategies have been employed by research groups to mask the antigenic sugar Galα1-3Gal for example, using extra copies of the blood group H antigen-specific α2-fucosyltransferase which converts the precursor of the Galα1-3Gal antigen into a non-antigenic form.[155] In early 2002, PPL Therapeutics announced the birth of pigs with only a *single* copy of the *α1, 3 galactosyl transferase (GT)* gene. Six months later, in July 2002, it announced it had produced the world's first double-gene 'knock-out' piglets, when four healthy pigs were born. A fifth piglet died of unknown causes shortly after birth. Other companies are working on constructing transgenic pigs expressing the human complement inhibitory protein, CD59 (hCD59) in combination with human membrane cofactor protein and human decay-accelerating factor.[156]

Even if strategies to eliminate HAR were successful, a process known as 'delayed xenograft rejection' (DXR) would result in the rejection of the organ in a matter of days. The mechanism of DXR is not well understood, and although endothelial cell activation is a key factor the triggers are not understood. Endothelial activation involves gene induction and protein synthesis that includes a shift to a procoagulant state, secretion of chemokines such as membrane cofactor protein-1, and induction of leukocyte adhesion molecules such as intercellular adhesion molecule-1, vascular cell adhesion molecule-1 and E-selectin.

9.4.1 Can we pretreat the recipient to prevent rejection?

A number of approaches have been trialled to reduce the preformed natural antibodies of the patient. These have used a combination of splenectomy to prevent antibody formation, plasmapheresis to remove circulating antibodies, organ perfusion, adsorption of anti-Galα1-3 Gal with soluble sugars, and directed immunosuppression. However, the complement cascade, because of its essential role, may make the best target.

There are a number of strategies available for therapeutic intervention in the complement cascade but there are serious limitations and currently there are few available data. Nevertheless, the side effects seen have included leukopoenia, thrombocytopoenia and infectious complications (such as cytomegalovirus infection and fungal infections[157]).

9.5 Problems with pigs[158]

9.5.1 Will pig hearts function in humans?

Even if the rejection problems are overcome it is still not certain whether a pig heart transplanted into a human will function normally and respond to host signals such as hormones. In fact it has been suggested that pig kidneys will be impossible to use because of gross biochemical differences between humans and pigs.

9.5.2 Xenozoonoses

Another important issue relating to the use of pigs as source animals for xenotransplants is ensuring that the animals are free from disease. This would require a highly controlled environment which at its most extreme could mean that the piglets were delivered by caesarean section directly into an incubator, the sow would be killed after the piglets were born and the piglets would be raised in isolation from other piglets. Such measures may be critical to the success of xenotransplantation, but there is a need to balance the need for effective disease control against adverse effects on the pigs.

Arguably the strongest argument against use of porcine organs is the worry of xenozoonoses, the transmission of infectious agents from one species to anther via the transplanted organ. It has been discovered that pigs harbour porcine endogenous retroviruses, PERVs, which can infect human cells *in vitro*. Multiple copies of retroviruses are integrated in the pig genome, which suggests that breeding clean pigs will be extremely difficult. Research has shown that diabetic mice treated with pancreatic cells derived from pigs could become infected with a pig virus – this was the first time that porcine endogenous virus had been shown to infect another animal. It has also been shown that pigs can act as vectors passing chicken flu to humans. Other potential viral pathogens include porcine cytomegalovirus (PCMV) and porcine lymphotrophic herpesvirus (PLHV). Porcine cytomegalovirus can be excluded from herds by sterile delivery as described, but the risks associated with PLHV remain unclear.[159] For these reasons many people view the use of animal organs as a potential risk not only to the recipient but also to the general public. How can we therefore make sure they are safe when there is no way to screen for unknown viruses?

There have also been fears concerning TSEs (transmissible spongiform encephelopathies) following the BSE (bovine spongiform encephopathy) crisis which led to new-variant CJD in humans in the UK.

On a more optimistic note, a study by Paradis in *Science*,[160] a retrospective study of 160 patients treated with pig skin grafts or pancreatic islet cells or who had had their blood extracorporeally perfused by pig organs in the preceding 12 years, showed no evidence of viral infection. The study included 36 immunosuppressed patients, all of whom were shown to be free of pig viruses. Interestingly, microchimerism (the presence of donor cells in the recipient) was seen in 23 patients for periods of up to 8.5 years.

The study adds fuel to the debate over whether xenotransplantation is safe for humans and should be contrasted with reports in *Nature Medicine* (1997, 3; 282–6) and Nature (1997, 389; 681), in which researchers warned of the possibility of pig viruses spreading to humans. No evidence of viral infection was found in a separate study, of human patients receiving pig islet xenotransplantation. Eighteen patients were monitored for up to 9 years for pig virus infection including pig endogenous retrovirus, pig cytomegalovirus, pig lymphotropic herpesvirus and pig circovirus type 2.[161]

9.5.3 Religious objections

Additional objection to porcine organs would be from religious groups who see the pig as unclean – some will not eat or touch pigs or material derived from them. The Pontifical Academy for Life[162] states the views of the Catholic Church: '*For a theological reflection that will help to formulate an ethical assessment on the practice of xenotransplantation, we do well to consider what the intention of the Creator was in bringing animals into existence. Since they are creatures, animals have their own specific value which man must recognize and respect. However, God placed them, together with the other nonhuman creatures, at the service of man, so that man could achieve his overall development also through them.*' However, the same paper goes on to say: '*A serious ethical commitment on the part of scientists should not neglect to explore therapeutic paths which may represent alternatives to xenotransplantation, such as seem to be promised by many recent discoveries in the field of genetics, as in a longer period the therapeutic use of adult stem cells.*'

In both Judaism and Islam[163] the raising and consumption of pigs is forbidden. However, xenotransplantation using pig organs is not regarded as eating pork. The holy texts of both Judaism and Islam allow exceptions to dietary laws in situations in which a human life is at risk.[164,165]

In Hinduism, the body must remain whole in order to pass into eternal life. However, under Hindu law individuals may donate their organs or accept an organ. In terms of xenotransplantation, with the exception of cows, which are sacred, there are no other laws on the use of animals to alleviate human suffering. Indeed, according the Shiva Purana, Lord Ganesha (Lord Shiva's son) received a xenograft of an elephant head after Lord Shiva accidentally severed his head.[166] As stated previously, Buddhism regards organ donation as a matter for an individual's conscience and the same applies to xenotransplantation.

9.5.4 Animal activists

Even though in the UK pigs are bred solely to be eaten, animal rights activists have also rejected the idea of xenotransplantation as it involves vivisection which they deem to be cruel to the animal.

9.6 Government legislation

In the UK and the USA, there have been moratoria on the use of pig heart transplants. A change in the legislation will require proof that xenozoonoses will not be a problem. Current reports suggest that individuals exposed to pig cells or pig proteins as part of a therapeutic strategy do not harbour pig viruses. There is still concern that infectious agents will take advantage of immunocompromised patients and possibly recombine with the host, producing novel infectious agents that could cause a public health problem. Officials at the Department of Health in the USA published revised guidelines in May 2000, reiterating that the FDA needs to keep a tight control over xenotransplantation. The guidelines suggest that 'no recipients of xeno-organs, their relatives, close contacts or partners should be allowed to donate blood or organs', that 'animals destined for use should be delivered into a sterile environment surgically, be housed in specialist facilities and be actively monitored for infections', and that 'patients must be counselled for long-term co-operation in surveillance'. There were US applications to conduct clinical trials and in the UK PPL and the Roslin Institute produced cloned pigs but then pulled out of xenotransplantation due to an increased interest in stem cell transplantation and therapeutic cloning which had been approved by the UK government. For further information see Chapter 17.

9.7 When will xenotransplantation start?

Cellular xenotransplantation is likely to offer the only hope for xenotransplantation in the foreseeable future and such clinical trials are already under way. However, even in this field the success rates in terms of both physiological function and graft survival remain poor. An example of one such trial is the potential use of porcine neural tissue for the treatment of degenerative diseases such as Parkinson's disease. The structural and functional properties of neural tissue across species are remarkably similar; however, to date the long-term survival of the grafts is very poor and clinical benefits are extremely modest. A good knowledge of the native biochemical environment of neural tissue and the processes of proliferation, migration and differentiation are therefore essential to the survival and function of xenogeneic tissue in the host brain. Studies are now being carried out, using various growth factors and inhibitors, to determine the optimal environment for neural graft survival.

Another example is the potential of animal pancreatic islets for use in the treatment of diabetes; this has the advantage that the compatibility of human and porcine insulin is already proven and there has been one clinical report of viable xenografted porcine cells containing insulin and glucagon although the insulin requirements of the recipient remained unchanged. For this to be successful there must be sufficient viable grafted cells to produce a clinical response and cells must be able to grow and develop normally in their new environment. Further work will be necessary in order to improve the functional properties of these cells in their new setting. With advances in stem cell therapy, xenogeneic stem cells are likely to be explored as a potential future therapy for a wide range of conditions.

9.8 Patient attitudes

A survey was undertaken in the UK by the British Kidney Association. They asked 850 patients known to them how they would respond and why to the offer of a xenograft; 663 (78%) were willing to receive pig kidney,

144 (17%) were not and 43 (5%) were unsure. The reasons against included religion and the special breeding of pigs for donation.[167]

9.9 Ethics

In terms of human rights, key ethical questions include 'On whom do we do the first test?', when the procedure involves, for example, removing a heart and not knowing if the replacement will work. We have to start with a seriously ill patient but one who is still alive and whose lifespan is unknown. For a heart transplant, failure means death, but we need to trial the therapy on a patient in order to determine the problems. In this kind of case, what is a successful outcome to the study? What justifies more transplants? When do we start and when do we stop? Human–human cardiac transplantation is not a cure; patients take immunosuppressive drugs for life and must be monitored constantly for infection, rejection and arteriopathy. Current 5-year success rates are estimated at 80% for kidneys and 77% for hearts. We need to contemplate how we will assess the success or failure of xenotransplantation and on whom it should be tested.

9.10 Alternatives to xenotransplants

We have known the risk factors for heart disease for many years; they include smoking, hypertension and poorly controlled cholesterol levels, to name but a few. We have tried strategies to improve cardiac health, but in the UK, for example, despite campaigns and education, there has been an increase in the number of women who smoke. Nevertheless, there are some techniques and technologies that could limit the number of transplants required, such as:

1. Preventive health and health maintenance programmes aimed at reducing the need for transplants.

2. Changes in the law to increase human organ donations, for example organ availability quadrupled in Austria when it changed to a 'presumed consent' law.

3. Clinical research to improve allotransplantation and rejection issues.

4. Improved surgical techniques to repair malformed or poorly functioning organs. In trials, 75% of patients who underwent ventricular
   remodelling, where a section of heart muscle is removed and reshaped,
   no longer required transplants.

With improved health education, rapid diagnosis and conventional therapy,
we could reduce the number of transplants required. The advent of stem
cell technology could enable us to repair organs rather than replace
them.

Points to consider

Ask your family and friends about their views on transplants and
   xenotransplants.
Do you or they carry a donor card?
Why do education policies fail to reduce the numbers of smokers?
How could we improve such health campaigns?

Notes

151 www.dlrm.org/resources/transplants.htm
152 www.cdc.gov/ncidod/eid/vol2no1/michler.htm
153 Ravelingien, A. and Braeckman, J. To the core of porcine matter: evaluating
     arguments against producing transgenic pigs. *Xenotransplantation*, 2004,
     11, 371–5
154 http://linkinghub.elsevier.com/retrieve/pii/S0022-5223(05)00692-6
155 www.fasebj.org/cgi/content/full/13/13/1762
156 www.blackwell-synergy.com/doi/abs/10.1111/j.1399-3089.2005.00209.x
157 www.emedicine.com/med/topic3715.htm
158 www.ishlt.org/PDF/pdf_xeno_guidelines.pdf
159 Patience, C. and Fishman, J.A. Xenotransplantation: infectious risk revisited.
     Am. J. Transplant., 2004, **4**, 1383–90
160 Paradis, K. *et al.* Search for cross-species transmission of porcine endogenous
     retrovirus in patients treated with living pig tissue *Science*, 1999, **285**,
     1236–41
161 Garkavenko, O. *et al.* Monitoring for presence of potentially xenotic viruses
     in recipients of pig islet xenotransplantation. *J. Clin. Microbiol.*, 2004, **42**,
     5353–6

162 www.ewtn.com/library/CURIA/PALXENOT.HTM
163 www.emedicine.com/med/topic3715.htm
164 Daar, A.S. Xenotransplantation and religion: the major monotheist religions. *Xeno.*, 1994, **2**, 61
165 Rosner, F. Pig organs for transplantation into humans: a Jewish view. *Mt. Sinai. J. Med.*, 1999, **66**, 314–19
166 http://en.wikipedia.org/wiki/Shiva_Purana
167 Kranenburg, L.W. *et al.* Reluctant acceptance of xenotransplantation in kidney patients on the waiting list for transplantation. *Soc. Sci. Med.*, 2005, **61**, 1828–34

10

Reproductive cloning

10.1 History

Key experiments involving the cloning of amphibians were carried out by John Gurdon in 1968.[168] He took a fully differentiated cell (either a keratinocyte or a nucleated red cell), removed the nucleus and placed that into an egg cell that had been enucleated by ultraviolet treatment. Some resulted in tadpoles genetically identical to the cell donor. His experiments were successful with nuclei from a limited range of differentiated cells and in only a few species and were important as they were the first to show that differentiation could be reversed, that a differentiated cell did not lose genetic material, but that genes are merely turned off. Interestingly, no tadpoles formed in this way went on to develop into adults. Cloning animals has been used in farming over the last 30 years. Basically a blastula is produced by IVF or harvested by lavage. The zona pellucida is destroyed and the blastula is split into small clusters of cells, each of which is implanted to produce offspring. Using this method it is possible to generate twins, quadruplets and up to 32 identical offspring. As long as the 'fate map' of the cells has not been laid down, any cell in the blastula is pluripotent and capable of producing all the cells needed by the embryo. This has been economically attractive in cows and horses where few offspring are naturally produced.

Negative aspects associated with this procedure include the phenomenon of large offspring syndrome (LOS).[169] The gene for the protein IGF2R

Molecular Therapeutics: 21st-century Medicine by Pamela Greenwell and Michelle McCulley.
© 2007 John Wiley & Sons, Ltd

(insulin-like growth factor 2 receptor) is imprinted, that is methylated, when inherited from the father but not from the mother. Thus, the protein is produced solely from the non-imprinted gene. In cloned animals methylation is often completely absent on the *IGF2R* gene and hence levels of IGF2R are very high. Since low levels of IGF2R help prevent the fetus from overgrowing the womb, the cloned animals grow much larger than normal. Nevertheless, there have been reports that, in humans, this would not be a problem, as IGF2R does not seem to be affected by imprinting. However, imprinting does occur in other genes and may provide us with different problems.

Dr Ian Wilmut and his team at the Roslin Institute were the first group to carry out cloning by transfer of nuclei from cultured embryonic stem cell lines; the experiments involved inducing cultured embryonic stem cells to enter the G_0 stage of the cell cycle. The experiment was a success as live lambs were produced. However, cloning from embryonic stem cells is of limited commercial use where the aim would be to produce a transgenic animal and then, once it was confirmed as a success, be able to clone from the fully differentiated cells of an adult animal. Wilmut persisted in this endeavour to harvest differentiated cells from live adult animals and use them for cloning rather than cultured stem cells, and in February 1997 'Dolly' the cloned sheep was announced as the result of a joint project between the Roslin Institute and PPL Therapeutics.[170] The cells from the udder of one adult sheep were taken and the nuclei were removed and placed in enucleated cells. After transfer to a surrogate mother, one sheep was born which was genetically identical to the udder cell donor.[171]

Table 10.1   Cloning methodology

- Take nuclei from donor cells such as mammary epithelium, fetal fibroblasts or embryos.
- Place nuclei into enucleated egg or fuse enucleated egg with the donor cell.
- Place egg into surrogate mother.
- Analyse progeny by DNA analysis.

In reality, we would take two different types of sheep for the experiment. The DNA donor, for example, could be a white-faced variety and the egg donor a black-faced animal. Offspring that were genuine clones would have the facial characteristics of the DNA donor, that is they would be black-faced.

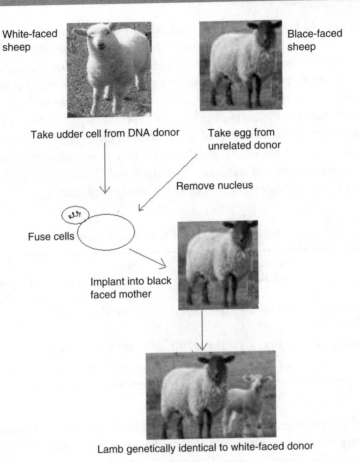

White-faced sheep

Blace-faced sheep

Take udder cell from DNA donor

Take egg from unrelated donor

Remove nucleus

Fuse cells

Implant into black faced mother

Lamb genetically identical to white-faced donor

Figure 10.1   Cloning sheep

10.2 Problems

The first hurdle that had to be overcome was the fact that in a fully differentiated adult nucleus only a small subset of genes are expressed, therefore the researchers needed to genetically re-program the adult nuclei. Essentially they did this by starving the adult cells until they became quiescent and entered into the G_0 phase of the cell cycle, this process allowing silent genes to be turned on and the process of nuclear transfer being facilitated by an electrical current which stimulates the new zygote into dividing. Whilst ground-breaking and a success in terms of demonstrating that cloning from a differentiated adult cell is possible, in its entirety the experiment cannot be described as a total success. The experiments carried out by Wilmut used nuclei from three distinct sources: mammary epithelial cells of a 6-year-old Finn Dorset ewe, fetal cells from a 26-day-old Black Welsh fetus and embryonic cells from a 9-day-old Poll Dorset embryo. Table 10.2 gives details of the overall success of the experiment. 'Fused couplets' represents the number of nuclear transfer experiments required to generate an apparently normal egg. 'Number recovered' refers to the number of these that were stable in the early stages of the experiment. 'Number of morulas' is a measure of the number of these manipulated eggs that progressed through the early stages of development *in vitro*. 'Number transferred' refers to the number of morulas implanted and 'Number of pregnancies' represents the number of surrogates that became pregnant.

Table 10.2   Production of cloned animals

Cell type	No. fused couplets	No. recovered	No. morulas	No. transferred	No. pregnancies
Mammary epithelium	277 (64%)	247 (89%)	29 (12%)	29	1
Fetal fibroblasts	172 (85%)	124 (87%)	47 (40%)	40	5
Embryos	385 (83%)	231 (85%)	126 (39%)	87	15

Of the experiments using mammary epithelial-cell-derived morulas there was only one that resulted in a live birth, which was Dolly the sheep. Of the fetal-cell-derived morulas there were three live births from five pregnancies; however, one fetus died soon after birth and there were four live births from fifteen pregnancies using morulas from embryonic cells. Overall, 62% of fetuses were lost in early pregnancy, which is high considering normal loss is estimated at 6%. Other interesting outcomes were that in the late-stage dead fetuses many gross anatomical problems were observed (e.g. abnormal livers). Similar results were later obtained by groups cloning pigs.

10.3 Why was their so much interest in Dolly?

The reason why Dolly represented a major scientific break-through was that she was the first animal to have been cloned from differentiated adult cells; this meant that now there is the opportunity for cloning adult animals of known characteristics. Since then, cloning by nuclear replacement using nuclei from adult cells has been successfully extended to mice and cattle. Dolly went on to reproduce but was euthanised on 14 February 2003 aged 6, suffering from a virus that caused a lung tumour. On autopsy no other gross abnormalities were found other than that she had arthritis which has been speculated to have been caused in part by her being overweight. A further interesting find was that her telomeres, the structures on the end of chromosomes that get shorter each time a cell divides, were 20% shorter than was normal for a sheep of her age. This led to speculation that Dolly's biological age might be equal to that of her own age and that of the donor ewe combined.

10.4 Was Dolly a lone example?

In Oregon, two rhesus monkeys were born as a result of clandestine cloning experiments that were only reported after Dolly had made the headlines. Dolly has since produced offspring. In Australia, cows have been cloned and there is interest in cloning rare wombat species. In Japan, mice have been cloned. Reports suggest that scientists have even cloned animals from cadavers (dead for about 30 minutes). Reports suggest that some cloned animals have died prematurely whereas others suggest that the cloning procedure has in fact rejuvenated the cloned offspring! Success rates have really not improved since Wilmut's first experiments.

10.5 Why is cloning useful?

The main drive and reasoning behind the experiments carried out at the Roslin Institute was to aid transgenic research. The idea was that it would enable researchers to clone sheep such as 'Polly', the transgenic sheep made at Roslin that secreted recombinant Factor IX in her milk. The ability to do this would mean it would be cheaper than current breeding programmes and also allow the rapid production of transgenic animals. There are also potential applications in animal breeding programmes, for example in beef cattle production, dairy farming and horse breeding, to produce a single stock of the appropriate sex (female). Cloning could create herds of cattle lacking the prion protein gene, or simply could be used in rare species preservation. A further, wider application of the cloning technology is to be able to better understand development and differentiation. Increasing our understanding in this area may enable us to generate matched stem cells and ultimately re-populate diseased organs, e.g. bone marrow, or be able to 'reprogram' a patient's own cells without the need for creating, and destroying, human embryos.

10.6 Is human cloning a reality?

After his success with Dolly, Wilmut was reported to have said 'Human clones will be a possibility in less than two years.' However, there have been no reports of successful human cloning, and there are unlikely to be any, especially as cloning humans is banned in the UK and the USA, along with many other countries in the world. In 1998 the Human Genetics Advisory Commission (HGAC) and the Human Fertilisation and Embryology Authority (HFEA) published a report after consultation with the public on the topic of human cloning.[172] The outcome was that there was very little support for reproductive cloning, with 80% of those consulted thinking that it was an ethically unacceptable procedure. Safety was seen as an ethical issue, particularly the wastage of human eggs that would need to be involved. The high risks of miscarriage and congenital malformation alone would exclude any realistic prospect of reproductive cloning. The use of reproductive cloning to relieve infertility is also likely to be ethically unacceptable for human use in the foreseeable future because of the resulting unbalanced genetic relationship.

The creation of an embryo through cell nuclear replacement for subsequent implantation into a woman's womb to produce a baby cannot take place in the UK. This is the technique most commonly referred to as cloning. A licence from the HFEA would be required for the creation of human embryos using this method of cloning. The HFEA has made it clear that as a matter of policy it will not issue a licence for treatment involving the use of embryos created in this way.

The uses of animal reproductive cloning, including the multiplication of animals with desirable characteristics and the creation of animals with new characteristics by genetic targeting, are not considered in reference 172. The policy of the Ministry of Agriculture, Fisheries and Foods (now DEFRA, Department for Environment, Food and Rural Affairs) on the cloning of farm animals was guided by the Report of the Committee to Consider the Ethical Implications of Emerging Technologies in the Breeding of Farm Animals (the Banner Committee) which was published in

1995. MAFF asked the Farm Animal Welfare Council to consider the implications of cloning for the welfare of farmed livestock.

10.7 Why can we not produce human clones that are identical?

The answer to this question is fundamentally one of practicalities; our current understanding of cloning and the lack of success associated with the process would make human cloning, at the very least, a risky business. Nuclear transfer technology is not easily applied in humans and will be not be safe; 277 'reconstructed eggs' were used to produce Dolly, and this required the surgical recovery of over 400 unfertilised eggs from donor ewes. Human IVF clinics recover an average of 5–10 human eggs per donor. This means you would need at least 40 volunteers for each prospective pregnancy. Another factor is that sheep are efficient at reproduction, more so than humans; a young fertile woman has a 1 in 3 chance of establishing a pregnancy after intercourse, even if the timing is right. Pregnancy rates in humans drop to 10–20% with IVF. The chances of successful pregnancy from a cloned human embryo are likely to be between a third and a tenth lower than in sheep. The cloning experiments at the Roslin Institute resulted in several lambs dying late in pregnancy or soon after birth, and some of these had developmental abnormalities. This has raised questions as to whether the 'reprogramming' of the transferred cells is 100% complete. Other worries include the possibility that when using adult cells the accumulated somatic mutations will be transferred with them, thus resulting in the clone having a shortened lifespan or increased chance of cancer.

Another factor that should be considered is that in cloning the genetic material is taken from the nucleus of a donor cell but the very early development of an embryo is reliant on the influence of the cytoplasm outside the nucleus that is derived from the egg donor. The cytoplasm contains DNA, RNA, proteins, mitochondria, ribosomes etc. And so one could argue that to make a true clone you would need the environmental factors to be the same as in the original pregnancy. So, to clone accurately you would take a nucleus from a female adult and place the nuclei in one of her mother's eggs that had been fertilised and enucleated. The only true clones we could raise would be females! Finally, the ethical point of view must be considered in terms of the interest of the prospective child.

10.8 So why clone humans?

In humans with maternal mitochondrial diseases cloning could allow the birth of a normal child. All mitochondria are inherited from the mother, so if the mother has a mitochondrial disease so will the offspring, but we could take a nucleus from the affected mother's fertilised egg and place it in an unaffected enucleated donor egg to yield an unaffected offspring. If a lesbian or homosexual couple wanted a genetically related child and were unwilling to use AID (artificial insemination by donor) or find a woman willing to bear their child than cloning could offer an answer. Similarly, if infertile couples wanted genetically related children, cloning would allow them to have a child genetically identical to one of them. Usually only one of a couple would be infertile and AID or surrogacy would be a more sensible move.

A company called Clonaid has offered reproductive human cloning on a worldwide basis to infertile couples, homosexual couples and people infected with the HIV virus as well as to families who have lost a family member. They also offered a range of services such as INSURACLONE™, OVULAID™ and CLONAPET™, and creating personal stem cells. Although they claimed to have produced at least five cloned babies, none has been analysed, or indeed seen.[173] Despite legislation, there have been several individual proponents of cloning who see cloning humans as a much more real possibility; these include Severino Antinori, an Italian fertility expert.

10.9 What are the ethical and moral problems?

We are unsure of the long-term problems associated with cloned humans. In animals, we have seen shortened lifespan and 'dopey cow syndrome'. The losses post-fertilisation are immense and it is unclear how eggs would be obtained, given that as many as 99% may be wasted. Equally, there are no compelling scientific reasons to clone humans and many of the fertility problems listed could be tackled by other, proven, technologies. We will never produce phenotypically identical copies of humans with all of their thoughts and experience. We cannot live again through clones.

We have no real idea of how long cloned humans would live or whether they would suffer long-term damage since we do not understand the effects of cloning on development. It has been suggested that we may jeopardise future generations since if more babies were born by cloning there would be less genetic diversity. This would be nonsense unless the majority of newborns were clones.

In terms of the religious aspects, most religious texts begin by a god making the first men and/or women. In Christianity Adam and Eve were made in God's own image and indeed Eve was made from Adam's rib! Is this a precedent for cloning?

Harold Varmus, Director of the NIH in the USA, said he found the idea of cloning experiments 'offensive' but he went on to say that if a couple who were infertile wanted a genetically related baby cloning may be a morally defensible way of achieving this!

The UK government has stated that work that creates cloned human beings should not and cannot be carried out. Tessa Jowell, Minister for Public Health, made the position clear: 'We regard the deliberate cloning of human individuals as ethically unacceptable.'

The President of the US, Bill Clinton, asked the US National Bioethics Advisory Commission (NBAC) to report on the ethics of such procedures. He asked the heads of executive departments and agencies that 'no federal funds shall be allocated for cloning of human beings'. On 9 June 1997, the NBAC concluded that nuclear replacement technology for the purposes of creating a human was unsafe, and they recommended a ban on research into human cloning. The proposed legislation has a five-year 'sunset clause' that allows review on the legislation. The Cloning Prohibition Bill 1997 was sent to Congress for consideration. Currently in the US there is a ban on both reproductive and therapeutic cloning although the latter is being flouted in some states. Up-to-date information can be obtained on the views of a number of states and their legislative policy.[174,175]

A number of international bodies have been given the remit to oversee the legislation. The EU draft Biotechnology Patents Directive forbids the issue of patents on work involving deliberate cloning of human beings and the Council of Europe Bioethics Convention formulated a protocol forbidding the cloning of human beings. Article 11 of UNESCO's Declaration on the Human Genome and Human Rights, adopted on 11 November 1997, states that 'Practices which are contrary to human dignity, such as reproductive cloning of human beings, shall not be permitted.' The 'Universal Declaration on the Human Genome and Human Rights' was published by UNESCO, November 1997.

Table 10.3   European legislation documents

Country	Legislation
Denmark	Act No. 503 on a Scientific Ethical Committee System and the Handling of Biomedical Research Projects (1992 Act No. 460) on Medically Assisted Procreation in connection with medical treatment, diagnosis and research (1997). This confirms the Danish Parliament's position, as of 25 January 1995, that treatment cannot be initiated in areas where a research ban already exists under the 1992 Act.
Germany	Federal Embryo Protection Act 1990. The creation of an embryo genetically identical to another embryo or fetus or any living or dead person is an offence.
Norway	Law No. 56 on the medical use of biotechnology 1994. Implicitly prohibiting embryo cloning.
Slovakia	1994 Health Care Law. Implicitly prohibiting embryo cloning. Spain Law No. 35/1988 on Assisted Reproduction Procedures. Explicitly prohibiting embryo and oocyte cloning with criminal sanctions.
Sweden	Law No. 115, 14 March 1991. Implicitly prohibiting embryo and oocyte cloning with criminal sanctions.
Switzerland	Federal Constitution. Legally binding, implicitly prohibiting embryo cloning. If adopted, the Federal Bill on Medically Assisted Procreation 1997 will explicitly prohibit embryo and oocyte cloning with criminal sanctions.

Table 10.4   UK and EU legislation

Country	Legislation
UK	HFE Act 1990 Schedule 2 paragraph 3 (2). HMSO
UK	Report of the Committee of Inquiry into Human Fertilisation and Embryology, HMSO, July 1984
UK	'The Cloning of Animals from Adult Cells', House of Commons Science and Technology Committee, Session 1996–97, Fifth Report (printed 18 March 1997), Vol. I
UK	'The Cloning of Animals from Adult Cells', Government Response to the Fifth Report of the House of Commons Select Committee on Science and Technology, Session 1996–97 (Cm 3815), Page 4, paragraph 17
UK	House of Commons Official Report, Parliamentary Debates (Hansard) 26 June 1997, Column 615
EU	European Parliament and Council Directive on the legal protection of biotechnological inventions COM (97) 446 final
EU	Council of Europe. Convention for the Protection of Human Rights and Dignity of the Human Being with regard to the Application of Biology and Medicine. Strasbourg: Council of Europe 1996 (ETS 164)

Points to consider

What is the value of cloning?
Would you agree to donate cells for this purpose?

Notes

168 Gurdon, J.B. Transplanted nuclei and cell differentiation. *Sci. Am.*, 1968, **219**, 24–35

169 McEvoy, T.G. *et al.* Consequences of manipulating gametes and embryos of ruminant species. *Reprod. Suppl.*, 2003, **61**, 167–82

170 www.roslin.ac.uk/publicInterest/cloningDiscussionPapers.php

171 Wilmut, I. *et al.* Viable offspring derived from fetal and adult mammalian cells. *Nature*, 1997, **385**, 810–13

172 www.advisorybodies.doh.gov.uk/hgac/papers/papers_c.htm

173 www.clonaid.com/news.php

174 www.ncsl.org/programs/health/genetics/rt-shcl.htm

175 www.ornl.gov/sci/techresources/Human_Genome/elsi/cloning.shtml

11

Stem cell therapy

11.1 The potency of cells

The following definitions are important to understand before discussing stem cells:

- Totipotent = able to form all cell types
- Pluripotent = able to form cells of a single lineage, e.g. bone marrow

Some researchers have suggested that we should change the nomenclature:

- Totipotent = newly fertilised egg
- Pluripotent = able to form most cell types, but not a whole embryo
- Multipotent = restricted to cell lineage

11.2 Cloning

Much of our knowledge of stem cells was derived from early experiments on cloning. We have already talked about Gurdon's work with toads and his discovery that, given the right conditions, nuclei from mature cells

Molecular Therapeutics: 21st-century Medicine by Pamela Greenwell and Michelle McCulley.
© 2007 John Wiley & Sons, Ltd

could be used in cloning experiments, suggesting that the fate of the cell depends not upon its genetic complement alone, but also on the cytoplasmic environment in which it finds itself. Hence, a nucleus in a keratinocyte has genes turned on appropriate to that cell type, but when removed and placed into an egg cell it re-expresses genes which are those of the egg. We have also discussed extending this work to mammals, for example Dolly the sheep. We also noted that eggs can be split to form multiple identical embryos, suggesting that, for a short time, all the cells in the early fertilised egg can go on to form the embryo proper, i.e. they are totipotent. Such 'natural' totipotency is confined to the early embryo.

11.3 Potency of stem cells[176]

It is clear that stem cells, if isolated, could provide us with an endless supply of totipotent or pluripotent cells. However, there has been much dispute about stem cells and their totipotency: their ability to form all cells, e.g. blastocysts or their pluripotency; their ability to form one type of cell lineage, e.g. haemopoeitic stem cells. If researchers could identify and grow totipotent cells in culture they could in theory supply endless numbers of uncommitted cells for transplant or they could differentiate the cells *in vitro* to yield committed cells for transplantation. Alternatively, a single organ could be repopulated with cells of one lineage using pluripotent cells.

11.4 Potential sources of stem cells

One of the major contentious issues has been the source of stem cells. We know that the best source would be the early embryo at the blastula stage, but this has led to worries about supply and ethics. We would also still need to match the donor and recipient.

11.5 Stem cells and therapeutic cloning

Cloning could, in theory, be used to create therapeutic material such as cells and organs. Such cloning has been termed 'therapeutic cloning' and is not intended to produce viable offspring, but simply embryonic stem cells. We are really talking about either producing a blastula from a cloned cell of the individual requiring therapy and then harvesting early stem cells that could be differentiated into tissues for transplant, or harvesting stem cells from viable embryos.

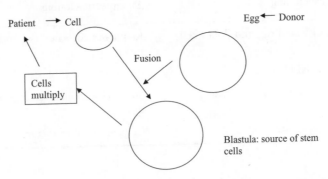

Figure 11.1   Therapeutic cloning

In theory, we could produce identical stem cells for the patient, but in practice the practicalities of the technology would be difficult to overcome. It is estimated that, in order to guarantee success, we would need to start with about 20 human eggs. It is difficult to imagine how these will be acquired in large enough numbers. A sick patient may not be able to wait the 12 or 18 months required to carry out the experiment. Additionally, a patient with a genetic disease will not profit from transplantation of his or her own stem cells as these too will carry the genetic mutation. In reality we will probably use this technique to produce lines of cells that can then be grown in bulk and matched in a similar way to conventional allografts. Researchers have tried to make chimeric embryos with the egg derived from a cow and the nucleus derived from human cells. In this case ethical issues should be limited since the embryo will not have the potential for full development and there would be no shortage of eggs. Fears have been raised that embryos that are viable will be killed to produce human tissue for transplantation. There are deep-seated fears about therapeutic cloning. We know that the nucleus supplies the genetic material, but the cytoplasm, which will have been derived from another individual, also plays an important role. Indeed, all the proteins, chemicals, ribosomes, mitochondria and RNA in the egg are derived from the cytoplasm of the egg. Will the cells produced then be compatible with the nucleus donor? Will anyone be tempted, despite legislation, to clone whole human beings? What regulations would we require to be absolutely sure that no scientist was tempted to take that extra step and, having made a cloned blastula, implant it and produce a human?

Table 11.1 A comparison of reproductive and therapeutic cloning

Reproductive cloning	Therapeutic cloning
Requires enucleated egg and donor cell fusion	Requires enucleated egg and donor cell fusion
Produces cloned blastula	Produces cloned blastula
Blastula is implanted in surrogate	Stem cells are harvested, grown and differentiated
Cloned offspring are born	Cells are used in therapy

Currently we are really only thinking about BMT since we cannot make complex organs *in vitro*. Nevertheless there have been reports in the literature about cloning 'organs'. Many scientists also believe that we will circumvent cloning as a source of embryonic stem cells and are researching into de-differentiation and programme switching to enable, for example, bone marrow stem cells to be used as stem cells for the brain and muscle.

11.6 Legislation and therapeutic cloning

Perhaps not surprisingly, the legislation on the use of embryonic material involves the Human Embryo and Fertilisation Authority. In 1998 the Human Genetics Advisory Committee and the HEFA published a joint report entitled *Cloning Issues in Reproduction and Medicine*. The UK government's response was to set up an expert advisory group to examine therapeutic cloning. This led to the publication of a discussion paper presented by the Nuffield Council of Bioethics to the Chief Medical Officer's (CMO) Expert Advisory Group on Therapeutic Cloning. In April 2000, the Nuffield Council of Bioethics published its report, *Stem Cell Therapy: the Ethical Issues*.

The CMO's report, *Stem Cell Research: Medical Progress with Responsibility*, was published in June 2000. In December 2000, the UK Parliament voted to amend the HFEA regulations to allow the use of embryos for stem cell harvesting and therapeutic cloning. The House of Lords supported this in January 2001. That report supported therapeutic cloning but not the production of humans from cloned material. It also suggested banning the production of chimeric clones. In 2004 a special licensing committee of the UK Human Fertilisation and Embryology Authority met to decide whether to grant a licence to scientists at the Centre for Life at Newcastle University. This was approved, and was the first licence in the UK and Europe to allow research into therapeutic cloning, focusing on the treatment of diabetes. Many groups have come to the conclusion that therapeutic cloning will benefit man and is necessary since it is more likely to be successful in the short term than is the production of stem cells from adults. Nevertheless, it is a brave politician who will stand up and say they are in favour of 'killing embryos' or 'harvesting material from aborted fetuses' – areas that evoke strong public feeling!

In the USA, the National Bioethics Committee published in 1999 its report, *Ethical Issues in Human Stem Cell Research*. Federal funding is only available for work on cell lines established prior to August 2001, but different states are now developing different legislation. In the US in 2003, the House of Representatives passed by 241 to 155, the Human Cloning Prohibition Act (HR 534). This bill, backed by President Bush, bans the creation of human embryos by cloning. The House decisively rejected (by 231 votes to 174) a competing proposal that would have allowed and encouraged the creation of human embryos by cloning for therapeutic purposes, while banning the use of any cloned embryo to 'initiate a pregnancy'.

Legislation varies from country to country, and is frequently dependent on religion. However, the area of cloning has been rocked by a scandal that may result in stricter legislation worldwide.[177] The Korean scientist Hwang Woo-Suk, who had been involved with the production of the first cloned dog, was involved in harvesting embryonic cells from blastocysts. He claimed[178] to have made a series of cell lines using the stem cell nuclear transfer (SCNT) technique. In early 2006, it was revealed that the egg donors required for the technique were in fact females working within his laboratory. Further investigation showed that his so-called stem cells were in fact the result of parthenogenesis, the chemically induced division of unfertilised ova, and not the product of SCNT, therapeutic cloning.

Table 11.2   Revealing the fraud

Feb. 2004	Hwang Woo-Suk reports that the group has created 30 cloned human embryos
May 2005	Group reports they have isolated stem cell lines from skin cells of 11 individuals
Nov. 2005	Following claims that he used eggs from his staff, Hwang apologises
15 Dec. 2005	Claims made that stem cell research was faked
23 Dec. 2005	Academic panel reports results were 'intentionally fabricated'

There are now real fears that opponents of therapeutic cloning and embryo research will use this fraud to make the point that scientists cannot be trusted. If they break the rules, how can we be sure they would not indulge in reproductive cloning?

11.7 Other sources of stem cells

Embryos and aborted fetuses could also be used as a source of stem cells, but we have already talked about the issues of fetal material used in transplant. In the UK, the current problems (Alder Hey inquiry, Tissue Act) associated with the use and storage of children's organs will probably affect this area as well.[179]

The use of material from aborted fetuses is regulated in the UK by the *Review of the Guidance on Research Use of Fetuses and Fetal Material*, 1989.

This whole area is a political and ethical minefield. A less controversial source of pluripotent cells would be placenta and cord blood, although there is dispute about the ownership. Should a child's cord blood stem cells be kept for his or her use in later life, or used for someone else? Re-programming adult cells is now under way, harvesting stem cells and then using chemicals, growth factors and hormones to de-differentiate and re-differentiate the cells. As discussed in earlier chapters, questions have been raised as to the safety of administration of growth factors to 'normal' donors.

Table 11.3   Potential sources of stem cells

- Blastulas
- Cloned blastulas
- Chimeric cloned blastulas
- Embryos
- Germ cells or organs of aborted fetuses
- Placenta
- Cord blood
- Adults
- Re-programmed adult cells

11.8 What can be done?

So far, we have managed to harvest cells from human embryos, fetuses and adult marrow, peripheral blood and skin. Most of the work analysing such material has been carried out in animal models. Interestingly all the government reports[180] stress that we are not trying to make complex organs, rather we are trying to produce cells for re-population.

11.9 Experiments on embryonic cells

In the UK, until December 2000, no work was carried out in the area of harvesting stem cells from blastulas. However, in the USA, privately funded laboratories have isolated 5 stem cell lines from 14 blastulas from 36 embryos. It was, however, important to determine whether these cell lines could be stored and still differentiate, since the aim was to establish typed cell lines suitable for establishing a stem cell bank. In 2000, researchers in Singapore and Australia were able to verify that stored stem cell lines were still viable and could still differentiate, in this case to form neurons. Current research is directed at ensuring that these cell lines can be kept indefinitely without affecting their potential to differentiate.

11.10 Experiments on fetal tissue and cord blood

Stem cells may also be derived from fetal tissue[181] at abortion or from placenta or cord blood[182] at birth. Tissues that are rich in stem cells, e.g. liver, can be grown successfully in the laboratory. In general, the cells are pluripotent and lineage-specific. Cells harvested in this way cannot be used without prior thought about rejection. In the USA, they are saving cord blood for future use by its donor.

11.11 Stem cells from adult tissues[183]

The ideal solution would be to isolate stem cells from the adult and then cause the cells to differentiate along a particular lineage. In this case stem cells could be taken from the patient to minimise rejection. We do, however, have problems associated with harvesting stem cells, apart from bone marrow, skin and blood: other sources are either difficult to obtain or the stem cells within the tissues have not been identified.

Adult tissues have been used as a source of stem cells for some years; in the main we are limited to the use of cells from bone marrow, blood, brain, skin and liver. The major problems are:

(a) the isolation of stem cells and

(b) their growth and differentiation in culture.

Nevertheless, haemopoeitic stem cells have been used in the treatment of leukaemic patients for over 10 years and this mode of therapy is probably the most favoured, resulting in shorter 'in-hospital' time. These are pluripotent stem cells able to form all cells of the haemopoeitic lineage. Researchers have, *in vitro*, been able to produce blood, liver, muscle and brain cells from human bone-marrow stem cells and mouse neuronal stem cells have been differentiated to form cells of the bone marrow.[184]

11.12 Safety and technical problems

A number of questions remain to be answered:

1. Can we really be sure that stem cells we differentiate in culture are safe for use?

2. Will we need to do animal tests and monitor phase I human trials very closely?

3. Are we absolutely sure that even autologous transplants are completely rejection-free?

4. Will there be problems if the transplanted material is a mixture of differentiated and undifferentiated cells?

5. Would transplant cells age more rapidly?

6. Would cells accumulate mutations during their growth and differentiation *in vitro*?

7. Are we certain we can supply enough stem cells to the patient for therapeutic effect?

8. Are the timescales too great to contemplate autologous harvesting and differentiation or therapeutic cloning?

9. Could we generate stem cells that could truly emulate the capacity of newly fertilised eggs?

Once again the only way of working out where the problems lie will be to do the experiments, in animal models and human volunteers. Only then will we be able to address the questions posed.

11.13 Perceived scope of therapy

So will stem cells be useful? Table 11.4 outlines some of the potential uses. You should note that in no case are we talking about making organs, we are simply talking about supplying extra cells.

Table 11.4   Scope of stem cell transplantation

Replacing lost nerve cells in patients with Parkinsonia and Alzheimer's disease
Replacing pancreatic islet cells in diabetics
Changing the outcome of spinal cord injuries
Replacing nerve tissue in multiple sclerosis
Replacing damaged heart cells in congestive heart failure
Replacing bone cells in osteoporosis
Replacing liver cells in hepatitis and cirrhosis

11.14 Clinical trials of stem cell therapy

There are many reported trials of haemopoeitic stem cell therapy. These are, not surprisingly, fairly successful, having been the first form of stem cell therapy developed. Indeed, in many hospitals it is routine to mobilise stem cells for allogeneic and autologous transplantation.

There are web sites (www.hopkinskimmelcancercenter.org/clinicaltrials/index.cfm and www.mskcc.org/15480) full of clinical trial data concerning stem cell therapies. However, published trials are harder to find.[185] There are a number of reports of stem cells being used to repair ischaemic hearts with some, though limited, success.[186,187]

Another area of interest, though again with little success, is muscle-derived stem cells for use in muscle-wasting diseases.[188] There are currently controversies on the nature of muscle-cell precursors, but even if this were solved the major issue would be the transplantation of stem cells into all muscles to effect a cure. This will be a difficult problem.

In the treatment and repair of eye tissue, there have been reports of stem cells used to produce functional retinal cells[189] and of corneal tissue being

made from autologous oral mucosal stem cells.[190] The latter in particular
seems to hold great promise for those requiring corneal transplant. In
stroke patients improvement has been reported on the transplantation of
neuronal cells derived from stem cells. However, as yet there have been
few substantiated reports of, for example, restoration of function in para-
plegic patients.

11.15 What are the future prospects for stem cell research?

There is a need to reduce the 'hype' and public expectation of speedy cures
for all disease by the use of stem cells. There is much basic biology that is
required before stem cells will become mainstream therapy. We know that
they have potential, as witnessed by successes in haematological diseases
and in both cardiac and eye surgery. Scientists must take care as the public
now is developing a distrust of our claims and there are many fears of
abuse of the technology. The establishment of cell lines with multiple or
limited potential to form a cell bank would seem to be the ideal. However,
much work is still required to ensure the utility and safety of such cells.

Point to consider

Ask your family and friends about their views on the use of therapeutic
   cloning or the harvesting of fetal cells for transplant. Would they have
   a stem cell transplant if required?
For which diseases is organ transplantation a better option than stem cell
   therapy?
In the UK we are about to sanction the production of chimeric clones.
   What are your thoughts on this?

Notes

176 http://www.mrc.ac.uk/YourHealth/StemCellResearch/index.htm
177 Rusnak, A.J. and Chudley, A.E. Stem cell research: cloning, therapy and
     scientific fraud. *Clin. Genet.*, 2006, **70**, 302–5

178 Hwang, W.S. et al. Patient-specific embryonic stem cells derived from human SCNT blastocysts. *Science*, 2005, **308**, 1777–83. Erratum in: *Science*, 2005, **310**, 1769. Retraction in: Kennedy, D. *Science*, 2006, **311**, 335

179 English, V. and Sommerville, A. Presumed consent for transplantation: a dead issue after Alder Hey? *J. Med. Ethics*, 2003, **29**, 147–52

180 www.parliament.uk/post/pn174.pdf

181 http://linkinghub.elsevier.com/retrieve/pii/S1521-6934(04)00120-8

182 http://linkinghub.elsevier.com/retrieve/pii/S1521-6934(04)00123-3

183 http://jcp.bmjjournals.com/cgi/reprint/57/2/113

184 Romagnani, S. *et al.* Peripheral blood as a source of stem cells for regenerative medicine. *Expert Opin. Biol. Ther.*, 2006, **6**, 193–202

185 www.clinicaltrials.gov/ct/search?term=stem+cells&submit=Search

186 Winkler, J. *et al.* Embryonic stem cells for basic research and potential clinical applications in cardiology. *Biochim. Biophys. Acta*, 2005, **1740**, 240–8

187 Stamm, C. *et al.* Stem cell therapy for ischemic heart disease: beginning or end of the road? *Cell Transplant.*, 2006, **15**, Suppl. 1, S47–56

188 Lee-Pullen, T.F. and Grounds, M.D. Muscle-derived stem cells: implications for effective myoblast transfer therapy. *IUBMB Life*, 2005, **57**, 731–6

189 www.brps.org.uk/White/W_Stem_Cell_Therapy.html

190 Nishida, K. *et al.* Corneal reconstruction with tissue-engineered cell sheets composed of autologous oral mucosal epithelium. *N. Engl. J. Med.*, 2004, **351**, 1187–96

12

Gene augmentation therapy

12.1 Introduction

In the last three decades, advances in DNA technology have made gene therapy an attractive option for the treatment of human disease. However, as we shall see, gene therapy is theoretically simple but has been almost impossible to achieve in humans. This does not mean that the therapy will never be successful, but certainly much background work is required before gene therapy can be considered as a practicable therapeutic strategy. In essence, gene therapy aims to introduce DNA into the host cells and cause the host to produce a missing or required protein. Essentially it is similar to recombinant protein therapy except that the protein is made *in vivo* not *in vitro*. In theory this has three advantages:

1. As the protein is produced within the host cell, it can be targeted to organelles or membranes and hence we could treat disorders such as cystic fibrosis which are not amenable to protein replacement therapy.

2. As the host produces the protein itself we should be free of worries about incorrect protein processing.

3. We could effect a cure if we could treat stem cells (for example for adenosine deaminase deficiency).

Molecular Therapeutics: 21st-century Medicine by Pamela Greenwell and Michelle McCulley.

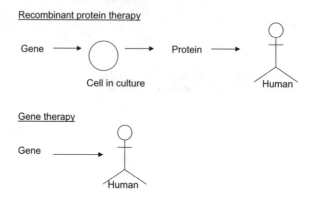

Figure 12.1   Comparison of recombinant protein and gene therapy

There are three types of gene therapy: somatic, *in utero* and germline. Somatic gene therapy involves treating cells which are not part of the reproductive system and achieving a cure within one patient but changes in their genome cannot be passed on to their offspring. *In utero* gene therapy involves gene therapy of fetuses whilst in the womb. There is some concern that the germline will be affected. Germline gene therapy by comparison involves manipulation of ova, sperm or eggs in such a way that any changes made in the genome to effect a cure could be passed on to the next generation. Currently *in utero* and germline gene therapy in humans are banned in the UK and the USA, although, as we have already discussed, this technique is used to produce transgenic animals. It is advisable to contemplate whether germline therapy should ever be used in humans, given the problems that have been seen in the production of transgenic animals by this method.

12.2 Strategy

Gene augmentation is the delivery of a gene to a cell with the view of augmenting gene function. This type of therapy may be useful in recessive disease where we know that in the presence of one good gene the patient will be cured. This is not a useful strategy for dominant diseases where we would need to remove the mutated gene – a technique well beyond the scope of researchers at present. Augmentation therapy could also be used to deliver genes encoding proteins that would be responsible for stimulating the immune system to treat cancer or infections.

Prerequisites for this type of therapy are delivery of the therapeutic gene into the cells and preferably into the chromosomes, targeting of the gene or gene expression into specific cells and finally production of correct levels of gene expression at the appropriate time. Unfortunately, these are all hurdles that have yet to be successfully overcome. We therefore need to explore these areas in more depth to understand why success has been limited.

12.2.1 DNA delivery

This is really the first problem and until we can successfully deliver DNA into cells augmentation therapy cannot be used. In theory there are many methods available for DNA delivery, some, for example calcium phosphate precipitation, electroporation and microinjection, are useful only *in vitro*. Electroporation, calcium phosphate precipitation and liposome delivery are too inefficient for safe delivery of therapeutic levels of DNA. The most popular methods of DNA delivery have involved viral vectors, although these do raise safety issues. Based on the idea that viruses naturally infect cells delivering their own DNA, scientists manipulated viral genomes to carry therapeutic genes. Viral vectors may integrate DNA into host chromosomes, can carry large amounts of DNA into cells and are very efficient. At first, two types of viruses were used, adenoviruses and retroviruses. Each had its advantages and problems.

The problems associated with viral vectors, which are discussed later in this chapter, were so severe that other groups began to develop non-viral vectors to achieve safe delivery of DNA. Liposomes were popular vectors but have proven to be relatively inefficient in terms of DNA delivery; they will not move out of the blood stream and realistically need to be conjugated to proteins such as growth factors to ensure delivery to the appropriate cells. We shall see that although liposomes promised much, they have, even with modifications, failed to deliver. So, what would be the characteristics of an ideal vector? Table 12.1 outlines the characteristics required. As yet we have no vector that fits all the criteria.

Table 12.1   Characteristics of the ideal vector

- a non-pathogenic virus
- capable of integrating DNA into the host genome at a single site
- able to carry large amounts of foreign DNA
- unable to cause an immune response in the host
- able to target cells specifically

NO SUCH VECTOR EXISTS

12.2.2 Viral vectors and gene therapy

Since there is no ideal vector we need to discuss the experiments carried out using conventional viral vectors for gene therapy. There are basically three types of viral vector commonly used in gene therapy: adenoviruses, retroviruses and adeno-associated viruses. Less commonly herpesvirus, poxvirus and vaccinia have also been used. We will confine ourselves to the most commonly used viruses and analyse their usefulness in gene delivery, for further information a website[191] examines all available viral vectors for use in humans. Many of the viral vectors we will study have been used very successfully in tissue culture cells and in animal models. However, often human-specific problems arise, many of which involve our immune systems.

12.2.2.1 Adenoviruses

Adenoviruses are non-integrating viruses that can infect dividing and non-dividing cells. They are a natural infectious agent affecting humans and have a natural tropism (affinity) for lungs. The viral genome contains genes whose removal will not harm the virus but will allow the incorporation of foreign DNA. Removal of the viral *E1* and *E3* genes enables us to incorporate up to 7.5 kb of human DNA. However, the function of the *E3* gene appears to be modulation of the host immune response. Hence removal of this gene may enhance the host immune response to the viral vector. The removal of the *E1* gene is important since this leads to the production of replication-incompetent viruses which can only infect the original target cell and its progeny and not spread to other areas of the patient or be released as an infectious agent into the environment. In order to make viruses which can infect the host cells we need to create our *E1*-negative virus with the inserted foreign DNA and grow it in a cell that contains a plasmid which can produce the E1 protein. The new viral particles are therefore packaged with the help of the added plasmid and can be harvested and used to infect the cells of choice. They cannot then infect other cells since, in the absence of the plasmid, no E1 products are available in the new host (in theory).

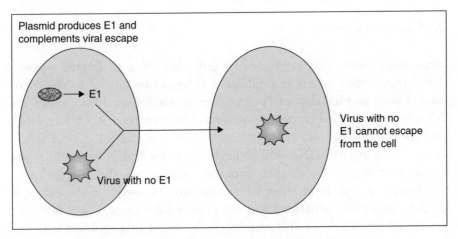

Figure 12.2   Replication-incompetent cells producing virus

In many ways, adenoviruses seem to be useful vectors. The fact that they do not integrate their DNA into the host, however, means that any response is transient since the DNA will be lost from the host cell. This transient expression is a real problem since, if we use adenoviral vectors, we will be required to give multiple injections to the patients. This in turn leads to priming of the immune system and production of a strong antibody response that then renders the treatment useless. In dogs treated for haemophilia using adenoviral vectors, the first round of gene therapy showed promise with good levels of Factor IX produced, but the second round led to a significant antibody response that rendered the therapy useless. So, the take-home message is that adenoviruses may be useful if a one-off delivery is required but, unless we can modify them, they will not be useful in gene therapy for inherited diseases. It is worthwhile to consider the problems in the use of adenoviruses, which include:

- transient expression leads to multiple injections and potential priming of the immune system

- adenovirus coat proteins are highly immunogenic and the patients develop strong antibody response to the virus.

A number of companies are trying to develop adenoviruses that contain no genes encoding potentially immunogenic proteins. However, some immunogenic cell surface proteins are also used in binding to the host target cells. So, reduction in immunogenicity results in reduced gene delivery.

12.2.2.2 Retroviruses

Retroviruses have also been used in gene therapy with limited success. The retroviruses used are non-infective. It would be unwise to use retroviruses such as HIV that are potent human pathogens. Retroviruses are RNA viruses that use reverse transcriptase to copy their RNA into DNA and then insert this into the host DNA. This is useful since, once incorporated, the inserted DNA will be faithfully replicated. If we could target pluripotent stem cells, all their progeny would carry the inserted gene, resulting in a long-term effect of the therapy. However, if the vector is used to infect differentiated cells, the effect of the therapy will last only for the lifetime of that cell. Immunogenicity is less of a problem with retroviral vectors, although immunological problems such as complement activation have been described. Retroviral vectors can be manipulated to carry up to 7kb (kilobases) of therapeutic DNA and can be synthesised as replication-incompetent particles. Production of high-titre viruses has been a problem, but this has been overcome by increasing efficiency of take-up by the cell after introduction of genes for stomatitis cell surface protein, which enables the retrovirus to bind to the cell more effectively.

Retroviruses, however, have limitations and disadvantages. One limitation is that, although the therapeutic gene is integrated, integration is random and could cause insertional mutagenesis. If the viruses infect a cell that contains other retroviruses, they could escape into the environment or infect inappropriate cells. However, possibly the greatest drawback in the use of retroviruses is that they will infect only dividing cells.

12.2.2.3 Adeno-associated viruses (AAV)

Because of the problems associated with adenoviruses and retroviruses, adeno-associated viruses (AAVs), which are parvoviruses (DNA viruses) not associated with human disease, have been investigated as potential vectors. The most commonly used is AAV2. The apparent advantages are that *in vivo* this virus integrates at a single site in the human genome on chromosome 19 and that integration causes no problems to the host. However, it has been shown that, once manipulated, the viruses integrate randomly. On the positive side they are also very efficient in their delivery of material to cells. However, AAVs need other viruses present to act as 'helper viruses' such as herpesvirus 6 and they have limited capacity with respect to uptake of foreign DNA. This clearly presents a problem.

12.2.3 Artificial viruses

Many groups are trying to make artificial viruses for therapeutic use.[192] Many of the experiments involve encasing DNA in viral coat proteins. The coat proteins facilitate delivery but also are the major stimulators of the immune response. We clearly outlined at the beginning of this chapter (see Figure 12.2) the features we would like in our ideal vector. As yet we cannot make such an artificial virus.

12.2.4 Non-viral delivery in gene therapy[193]

Since viral vectors have shown great promise in cell culture and early animal trials but proved difficult to use in humans, researchers have investigated non-viral delivery methods to attempt to overcome problems associated with the host immune system. Non-viral delivery methods include liposomes, particle bombardment and ligands.

12.2.4.1 Liposomes

Liposomes were the first candidates in the search for non-viral vectors. They are essentially small fatty globules that in culture will bind to cells and incorporate themselves into the cell membrane. If DNA is encased in or on the liposome, it is delivered to the cell. Many types of liposomes have been constructed, some with charge on their surfaces, for example cationic liposomes, some bearing protein such as growth factors which allow them to bind to specific cells.[194]

The major limitations in the use of liposomes are delivery to the cells, delivery to the nucleus, lack of efficiency of transport and toxicity at therapeutic levels. In fact, if we inject liposomes into the blood stream they will never escape: this clearly limits the *in vivo* therapeutic uses to easily accessible tissues such as lung. If we can deliver them directly to tissue, once the liposome has merged with the cell membrane and delivered the DNA that DNA must then find its way to the nucleus to be incorporated. In early trials, toxicity was not a problem, though efficiency was low. Later trials increased the amount of liposomes used. That increased efficiency of delivery and also increased the toxicity.

12.2.4.2 Particle bombardment

Particle bombardment has been described as a method of delivery of DNA in mice but has really not been popular in the treatment of humans. This technique involves coating particles, often gold, with DNA and 'firing' them at the area to be treated. In humans such technology may be useful for treating skin complaints but delivery to organs would be difficult. Since gene expression is transient, the patient would have to be re-treated and clearly the effect and fate of the carrier would need investigation. This is also a low-efficiency technique.

12.2.4.3 Ligands

Ligands are being used in combination with non-viral vector systems to improve DNA uptake and targeting. Non-viral vectors could be coated with specific ligands that would then aid targeting of the DNA to the required cells or organs. Such ligands include growth factors that will bind to cells expressing the relevant receptor, antibodies and lectins (sugar-binding proteins). We need to be mindful that these ligands will themselves be recombinant proteins and therefore their use may be problematical since they must also be human-compatible. It is also necessary to ensure that the ligand–vector complex is stable and resistant to enzyme degradation to allow targeted delivery.

Thus, gene therapy is not really successful in humans because of the lack of vector systems for gene delivery. However, once this is overcome, cell culture models suggest that DNA delivered to cells can be expressed and in the case of cystic fibrosis can restore function.

12.2.5 What tissues can we currently target?

You will notice that the types of organs we can target with gene therapy are similar to those we can replace by transplantation. The rationale for treating diseases with gene therapy is that we know that the patient will get better if that organ has a functional copy of the appropriate gene. However, gene therapy will not reverse organ damage.

12.2.5.1 Lungs

Lungs are attractive to the gene therapist since the cells are easily accessed using aerosols for gene delivery *in vivo*. However, since mainly surface cells are infected and these are sloughed off regularly, expression is transient. Adenoviral vectors have been used to treat lung diseases such as cystic fibrosis but side effects have been a problem. Liposomes have also been used but proved toxic at high dose. Retroviral vectors are inappropriate since the cells infected are not dividing.

Table 12.2   Diseases that could be amenable to gene therapy in the lungs

Disease	Characteristics
Cystic fibrosis	The most common genetic disorder in Caucasians and most pathology is associated with the lungs
α-1-anti-trypsin disease	Caused by the absence of an inhibitor of protease and most effects are seen in the lung, leading to emphysema and death

12.2.5.2 Bone marrow

Bone marrow is also an attractive target for the gene therapist since cells can be removed, manipulated and returned to the patient. The ability to isolate bone marrow stem cells offers the possibility of a long-term cure. Adeno- and retroviral vectors have been used. The first disease treated with gene therapy was adenosine deaminase deficiency, which can also be treated by transplant. Gaucher's and Batten's diseases are also amenable to treatment. Production of Factor VIII, which is naturally produced by the liver, could be induced in bone marrow since the final protein is required in blood. Although the haemoglobinopathies are common, gene

Table 12.3 Diseases that could be treated by gene augmentation in bone marrow

- adenosine deaminase deficiency is the model disease for gene therapy and was known to be curable by bone marrow transplant
- Gaucher's and Batten's diseases would be amenable to treatment
- introduction of genes for Factor VIII (normally expressed by liver) could cure haemophiliacs
- HIV could be a candidate since the virus is found in cells of the bone marrow
- haemoglobinopathies are relatively common *but* difficult to treat by gene therapy

Table 12.4 Diseases of the liver amenable to gene augmentation therapy

- Factor VIII deficiency
- α-1-anti-trypsin disease
- familial hypercholesterolaemia

therapy would be difficult as we would need to ensure the appropriate balance of α- and β-globin chains. Since we are currently finding it difficult to deliver the DNA, thoughts of treating disease where we need controlled gene expression are premature.

12.2.5.3 Liver

The liver is an interesting organ since small pieces may be removed surgically, manipulated and returned to the patient where they will settle down and grow. Experiments have involved *ex vivo* treatment of hepatocytes and their return to the liver via the hepatic portal vein, for example in the treatment of hypercholesterolaemia and *in vivo* infusion of retroviruses following a partial hepatectomy. *In vivo* techniques include partial hepatectomy, causing liver regeneration and simultaneous infusion of virus; in trials 2% of cells were infected.

12.2.5.4 Brain

The brain is difficult to treat using gene therapy since introduction of the vector is problematical. It would be difficult to penetrate the blood–brain barrier, and so direct injection is required into the brain. Additionally, since adult brain cells divide infrequently, retroviruses cannot be used to deliver genes to the brain. However, researchers have made use of this to treat brain cancers where cell division does occur. They have looked at delivering packaging cells, that is cells containing retroviral vectors, into brain, and direct injection of vectors into the brain, for the treatment of brain tumours. Direct injection causes cell death at the site of injection and localised infection of the vector. Vectors such as adenoviruses may be used and some researchers have worked with herpesvirus vector, which has a tropism for brain. Brain diseases such as Alzheimer's and Parkinson's may be amenable to such treatment. Researchers are also looking at ways to treat schizophrenia and CJD using this type of methodology.

Table 12.5   Disease candidates for gene therapy in the brain

---

- Alzheimer's disease
- Parkinson's disease
- schizophrenia
- CJD

12.2.5.5 Muscle

Muscle has been a focus for gene therapy since the disease Duchene's muscular dystrophy which affects young boys has no other feasible treatment. However, muscle is difficult to work with since it does not divide rapidly. We would also need very efficient delivery methods and would probably need to infect at least 5% of all muscle tissue to effect a cure. Workers have attempted to deliver DNA by direct injection and particle bombardment. Currently much attention is focused on the *ex vivo* treatment of isolated myoblasts which theoretically could be returned to the patient.

12.2.6 Targeting gene expression

Currently our major problem is DNA delivery; once this is solved our next problem will be cell targeting. If we cannot target the DNA to specific cells we can incorporate tissue-specific expression sequences which ensure that the introduced genes are only activated in the appropriate tissue. This is analogous to the targeting we described to facilitate the production of recombinant proteins in milk. However, the increase in the size of the insert DNA may itself prove to be a problem. This is an important issue since many genes are expressed in a tissue-specific manner, for example globin. The consequences of production of globin in other cell types are unknown.

12.2.7 Problems associated with augmentation therapy

In addition to problems of DNA delivery, targeting and expression, there are other more patient-related problems we should consider. The first of these is that gene therapy is unlikely to help anyone whose organs have been damaged by their disease. It is naïve to imagine that the addition of a single gene will miraculously completely cure a patient whose condition has led to end-stage organ failure, deposition of toxic products or destruction of large numbers of cells. Augmentation therapy is also really limited to the treatment of recessive diseases and the delivery of genes related to cancer and infection. We cannot replace genes – we can only add in more copies; currently treatments are considered experimental and potentially dangerous and are therefore carried out on terminally ill patients with no other therapeutic options available to them. These are not ideal subjects!

Since germline gene therapy is banned, we can only carry out somatic therapy. This may help the patient but they may then pass on their 'bad' genes to their offspring who in turn may need treatment. Even if germline therapy were used, there are immense problems associated with the technology and fears of eugenic practices.[195] We need to remember that we have much experience in the production of transgenic animals by germline gene therapy and are well aware of the problems that have arisen. We are not in a position where we would willingly undertake such therapeutic strategies on humans.

In utero gene therapy has been suggested whereby fetuses detected as abnormal in prenatal diagnosis are given gene therapy whilst in the womb.[196] The advantages are that the therapy targets stem cells and therefore would constitute a cure and that the child never develops symptoms. The downside is that, in work with sheep, the genes were also found in reproductive tissue. There are ethical problems associated with testing this therapy. It has been suggested that the therapy could be trialled on affected fetuses that were to be aborted. Therapy would be administered and a few weeks later the fetus aborted and the outcome determined. This is fraught with ethical problems. Common sense would suggest that in families at risk we should be offering PIGD (Chapter 2) rather than germline gene therapy.

12.2.8 Gene augmentation therapy vs. recombinant DNA therapy or transplantation

As we can see from Table 12.6, recombinant protein therapy, where appropriate, is probably a better therapeutic option at the moment than is gene therapy. It cannot, however, offer a cure and the patients require costly long-term therapy. There are problems associated with immune response to poorly processed proteins and the production of antibodies to proteins that the patient has never made (that is, the recombinant proteins which are seen as non-self). This is still a problem with gene therapy and there are a number of researchers investigating strategies to produce proteins lacking the most immunogenic regions.

Compared to transplantation, gene therapy currently does not appear to be a valid option, however; organ shortage and problems with organ matching and disease transmission limit the scope of transplantation. Organ transplantation is costly and dangerous and patients are on continual immunosuppression, which in later years predisposes them to cancer. One of the major advantages of organ transplantation is that it delivers new genes and a new fully functional organ whereas gene therapy cannot reverse severe tissue or organ damage. The genes in a new organ are under appropriate control and gene expression is fully controlled.

Table 12.6   A comparison of recombinant protein therapy vs. gene therapy

Recombinant protein	Gene therapy
Regular treatment required	Could give a one-off cure
Does not cure the disease, treats the symptoms	If successful, could cure the disease
Immune response can be seen to incorrectly processed proteins	Protein processing is correct
Immune problems	Immune problems
No viral problems	Viruses used
In vitro protein production	*In vivo* protein production

Table 12.7   A comparison of organ transplants vs. gene therapy

Transplants	Gene therapy
Validated	Experimental
Cure	If successful, could cure the disease
Controlled gene expression	Expression a problem
Treat terminally ill patients and reverse symptoms	Cannot reverse damage
Relatively cheap and well established	Expensive
Requires trained personnel	Requires skilled scientists

Table 12.8   Criteria for gene therapy

- patient has life-threatening condition
- no other treatment is successful
- gene has been cloned
- target organ is identified and accessible
- disease can be righted by the addition of a gene
- minimal gene expression will give improvement

12.2.9 Current criteria for the use of gene therapy

Since gene therapy is still experimental there are strict guidelines as to the types of diseases and patients to be treated. Currently only those with life-threatening conditions should be considered for gene therapy, given the paucity of safety data. All other therapeutic options should have been considered and proven unsuitable or unsuccessful. Recombinant protein therapy, if available, is currently a safer option, and even though transplantation is dangerous it is a proven therapeutic strategy and, if appropriate, would be offered before gene therapy.

Obviously we can only treat diseases where the gene has been cloned, the target organ has been identified and we have a strategy for delivering the DNA. Cell culture and animal studies must have been carried out to ensure that the addition of the gene will rectify the condition even if the expression of the gene is minimal. This last point is important since we cannot guarantee how much expression we can obtain and in the main we are treating diseases where the actual amount of product is not important and where less than 10% of normal protein expression will effect a cure.

12.2.10 The bystander effect

The bystander effect is difficult to explain as in many aspects it is illogical; nevertheless, it is a fact that during gene therapy it has been observed that although limited numbers of cells have taken up the augmenting gene, many more are affected by the therapy and return to normality. This has been seen in the trials of therapies for CF. It is more understandable in gene therapy for cancer where induction of a few cells enabled killing of cells nearby by production of permeable toxic substances.

12.2.11 Candidates for gene therapy

Ideally we should only work with diseases in which minimal expression of the introduced gene will improve the condition of the patient. For example, in adenosine deaminase deficiency (ADA) a 'cure' can be effected by successful treatment of about 10% of the cells. The target cells should also be easily accessible in tissues such as bone marrow or lungs, as we find it difficult to target expression. Gene therapy for CF has been well investigated as the target organ, the lung, is relatively easy to access. Table 12.9 lists those diseases on which we would like to use gene therapy.

Table 12.9   Candidates for gene therapy

Disease	Gene	Population frequency
Cystic fibrosis	CFTR	1 in 10,000
Pituitary dwarfism	hGH	2 in 100,000
α-1-Anti-trypsin disease	α-1-anti-trypsin	2 in 10,000
Hypercholesterolaemia	LDL receptor	1 in 1,000,000
Lesch–Nyhan	HPRT	2 in 10,000
Thalassaemia	Globin	2 in 1,000
Sickle cell disease	β-globin	2 in 1,000
Haemophilia A	Factor VIII	1 in 10,000
Haemophilia B	Factor IX	3 in 100,000
Gaucher's disease	Gluco-cerebrosidase	4 in 10,000
Glycogen storage disease	Glycosidases	2 in 100,000
Phenylketonuria	Phenylalanine hydroxylase	5 in 10,000
Severe combined immune deficiency (SCID)	Adenosine deaminase	1 in 100,000
Duchene's muscular dystrophy	Dystrophin	2 in 10,000

However, some, such as thalassaemia and sickle cell disease, are problematical as they require controlled expression of the gene product. There has been some progress with the clotting disorders, though it is important here that there is not over-expression of factors, which could be fatal. We should note the frequencies of these diseases. They are rare and there has been much speculation as to the cost-effectiveness of the development of gene therapy programmes for what is essentially a handful of patients. On the other hand, for many diseases there are no alternative therapies and gene therapy offers the only likely treatment.

Points to consider

What diseases should we try and treat using gene therapy?

Given the low numbers of patients affected, are we wasting resources in this type of research that could be better spent elsewhere?

Read http://bmj.bmjjournals.com/cgi/content/extract/330/7482/79. How do you interpret the risk?

Notes

191 www.medscimonit.com/pub/vol_11/no_4/6257.pdf

192 Mastrobattista, E. et al. Nonviral gene delivery systems: from simple transfection agents to artificial viruses. *Drug Discovery Today: Technologies*, 2005, **2**, 103–9

193 Louise, C. Nonviral vectors. *Methods Mol. Biol.*, 2006, **333**, 201–26

194 Kostarelos, K. and Miller, A.D. What role can chemistry play in cationic liposome-based gene therapy research today? *Adv. Genet.*, 2005 **53**, 71–118

195 Smith, K.R. Gene therapy: theoretical and bioethical concepts. *Arch. Med. Res.*, 2003, **34**, 247–68

196 Waddington, S.N. *et al. In utero* gene therapy: current challenges and perspectives. *Mol. Ther.*, 2005, **11**, 661–76

13

Gene therapy trials for inherited diseases

13.1 Introduction

A number of gene therapy trials have taken place over the last 30 years. The early trials were completely unsuccessful because of a lack of under-standing of gene manipulation. Early workers tried to introduce Shope viruses, expressing arginase into arginase-deficient patients and therapeutic genes into patients suffering from xeroderma pigmentosa without thera-peutic effect. Munyon[197] managed to achieve expression of the *thymidine kinase* gene in cultured cells after cell transfection. In the 1980s work began on germline gene therapy with ground-breaking experiments which led to the production of transgenic animals. Experiments showed that injection of material into eggs had no detrimental effect and Palminter[198] went on to give growth hormone genes to dwarf mice inducing growth. Sadly this was uncontrolled growth leading to giant mice. A salutary lesson perhaps.

In 1987 French Anderson began his pioneering work on gene therapy in humans. The first disease treated with gene augmentation therapy was adenosine deaminase (ADA) deficiency, a disease that results in the birth of children with no effective immune system, the so-called 'bubble babies'. Treatment at that time involved BMT or recombinant protein therapy with PEG–ADA (polyethylene glycol linked to adenosine deaminase). French Anderson's original experiments use mouse leukaemia virus to deliver the gene to differentiated white cells of two young female patients. Since that

Molecular Therapeutics: 21st-century Medicine by Pamela Greenwell and Michelle McCulley.
© 2007 John Wiley & Sons, Ltd

time, hundreds of patients have taken part in gene augmentation therapy trials; most of them have derived little or no benefit. Gene therapy of this type is still experimental.

Table 13.1   Gene therapy history

1970s
- Terheggen and Rogers injected Shope virus which produces skin papillomas in rabbits and induces a virus-coded arginase into arginase-deficient patient – no effect
- Clever used viruses to treat xeroderma pigmentosa – no effect
- Munyon successfully transferred the *thymidine kinase* (TK) gene into cultured cells

1980s
- Gordon and Ruddle microinjected mouse embryos and showed this had no adverse effect
- Cline inserted *β-globin* gene into patients with β-thalassaemia – no results
- Palminter injected growth hormone gene into dwarf mice – giant animals were produced

1987 French Anderson carried out first sanctioned gene trials. Used mouse leukaemia virus (retrovirus) to infect bone marrow cells with the adenosine deaminase gene *ex vivo* and re-injected.

By June 2005, there had been 695 sanctioned gene therapy trials in the USA. These included 42 for infectious disease, 60 for monogenic diseases, 84 for a range of disorders as diverse as ulcers and arthritis, 459 for cancer.[199] Many diseases have been treated with retroviral vectors; these include ADA deficiency, Gaucher's disease and hypercholesterolaemia.

13.2 Examples of disease treated with retroviral gene therapy

13.2.1 Severe combined immunodeficiency

There are a number of variants of this disorder. Autosomal recessive SCID occurs in about 16% of affected individuals whereas X-linked SCID is found in 46%. ADA deficiency causes severe combined immune deficiency; children homozygous for this recessive disorder have no innate immunity and develop infections readily and die. The current treatments involve PEG–ADA (an enzyme replacement) or bone marrow transplant. However, for many affected children there are no suitable bone marrow donors and in some cases the PEG–ADA treatment is not well tolerated.

Table 13.2 Conventional treatments for ADA deficiency

Allogeneic bone marrow transplant
HLA identical 73% survival
Haploidentical 23.5% survival
PEG–ADA therapy
80% had protective immunity
20% had minimal improvement
~50% needed Ig supplements
10% develop antibodies to the enzyme
Overall survival 74%

Although a rare disease, ADA deficiency was considered as a good candidate for gene therapy since it was curable by BMT and in culture, provided about 10% of cells took up the gene, expression was seen at therapeutic levels. The *ADA* gene had been cloned and it was readily inserted and expressed in white cells. In the first human trials in 1991, peripheral white cells were infected rather than stem cells. The effect of therapy was therefore transient even though retroviral vectors had been used. Later trials in the USA and Italy utilised stem cells for long-lasting therapy. The effectiveness of the gene therapy has been questioned, as all children treated still received PEG–ADA. These children are still alive, although whether it is the gene therapy or the PEG–ADA effecting a cure has been difficult to assess, though reports suggest that children were alive and well 12 years post-treatment.[200] PEG–ADA is a recombinant protein complexed with an inert chemical and it has been used in over 150 patients worldwide either as an alternative to mismatched transplant or as a stabilising measure prior to transplant. It is reported that about two-thirds of patients treated with PEG–ADA have survived, with the majority of patients showing good clinical improvement.[201] The first UK trial for SCID involved one patient who is still alive with no evidence that the gene is still present.

Table 13.3   Gene therapy trials for ADA deficiency

Target cells	Country	Year	Number of patients
T-cells	USA	1990	2
T-cells + stem cells	Italy	1992	3
Cord stem cells	USA	1991	3
BM + stem cells	France/UK/Netherlands	1993	3
T-cells	Japan	1995	1
BM + stem cells	Italy	2001	6
BM + stem cells	USA	2001	4
BM + stem cells	Japan	2003	2
BM + stem cells	UK	2003	1

Table 13.4   Summary of a gene therapy trial for ADA using HSC[203]

- Patients tested: 22
- Patients in whom gene was detected for more than 30 months: 8
- Patients who showed clinical benefit: 7
- Patients with lymphoproliferative complications: none

In 2004, Aiuti[202] reported gene transfer in stem cells resulting in immune and metabolic correction of ADA-deficient SCID with clinical benefit. Aiuti noted that gene therapy was effective in the absence of enzyme-replacement therapy. This worker stated that the use of low-intensity conditioning improved the engraftment of gene-corrected cells. In their opinion, if results are confirmed in the long term, a single treatment may be sufficient to cure the patient, at costs similar to those of stem cell transplantation. Aiuti's current research agenda includes improvement of vector design and gene transfer protocol and identification of the optimal cell dose and conditioning regimen. If the results bear scrutiny, then similar gene transfer approaches could be used to treat other inherited or acquired severe disorders of the haematopoietic systems. However, in order to assess long-term safety and efficacy of gene therapy it will be necessary to extend the observation time and the number of patients enrolled.

However, in this trial there was retroviral integration similar to that seen in the trials with X-linked SCID. In the Necker Hospital in France, 9 boys were being treated in an X-linked SCID gene therapy trial. One developed a leukaemia. In the original reports, researchers said this was unfortunate but should not stop trials. The child had some predisposition to leukaemia and this was thought to have precipitated the disease. Thus, despite knowing the problems associated with retroviral vectors and cancer, researchers did not immediately connect the treatment with the leukaemia. A few weeks later, however, another child in the trial was found to have the same mutation in the *LMO-2* gene, brought about by the gene therapy. At that stage, the hospital stopped recruiting into the trial and trials were stopped in the UK and the USA. They have been restarted with the acceptance that, although there is a risk of leukaemia or cancer, as long as the benefits to the patient outweigh the risks then the therapy should continue.

In 2005, a report for the NIH Recombinant DNA Advisory Committee (RAC) in the USA[204] stated that the majority of children treated for X-linked SCID had benefited from this treatment. However, of the 9 boys treated in the Necker trial, 3 had developed leukaemia and 1 had died as result of this. The writers of the RAC report concluded not only that the gene therapy was the cause of the problem but also that it was an inherent risk in this type of therapeutic strategy. They concluded that in the future gene therapy should only be carried out on those with no other treatment options, for example BMT or stem cell transplant and that each case should be reviewed on its own merit. Interestingly, the article states: '*Some subjects in gene therapy for non-X-linked SCID have achieved mild to moderate improvement.*' This is certainly not as positive as reports from those working on the disease.

13.2.2 Hypercholesterolaemia

In some cases of this disease, the defect involves the lack of LDL (low-density lipoprotein) receptors which help absorb cholesterol. Normally the patients are treated with cholesterol sequestration by drug therapy. The original gene therapy trial was performed on one patient, a 29-year-old female who did not respond to drug therapy. This patient had a family history of disease with a high percentage of her close family dying before they reached 30 years. This patient had extremely high levels of cholesterol and was at grave risk of thrombosis. She therefore agreed to take part in a gene therapy procedure in which a lobe of her liver was removed, the hepatocytes were treated with the gene for low-density lipoprotein receptors, and the transduced cells were returned via her hepatic portal vein and migrated back to the liver.

Table 13.5   Gene therapy protocol for hypercholesterolaemia

- remove left lobe of liver
- separate cells with collagenase
- add virus with normal *LDL receptor* gene
- infuse cells into the hepatic portal vein
- cells settle in liver capillaries and express receptors

Table 13.6   Current treatment strategies for CF

- DNase I aerosol to decrease mucus viscosity
- Antibiotics to prevent infection
- Physiotherapy to mobilise mucus
- Digestive enzymes to enable digestion
- Intracytoplasmic sperm injection to overcome blockage of the vas deferens
- Lung transplants to replace damaged organ

The results were very encouraging. The patient's baseline level of lipids dropped to a level that made her responsive to the drug-based therapy. Since that time a handful of other patients have been treated for this condition and all are doing well. Prolonged reductions in LDL cholesterol were seen in three of the five patients, and this procedure was free of any major side effects.

This study demonstrated the feasibility of *ex vivo* therapy.[205] However, there are limitations to the numbers of autologous hepatocytes that can easily be harvested and manipulated. Advances in the propagation of hepatocytes, for example, have made this technique more useful. *In vivo* approaches to genetic manipulation involve the transfer of genes to target tissue by either systemic administration or direct injection: *in vivo* therapy has been applied to ornithine transcarbamylase (OTC) deficiency.

13.3 Cystic fibrosis

CF, the most common recessive disease seen in Caucasians (1 in 20 are carriers), is caused by a mutation in the *CFTR* gene. Patients homozygous for the CF mutation suffer problems associated with the production of thick and sticky mucus. Their lung function is compromised, they require pancreatic enzymes to digest their food and many are infertile. Current therapies include physiotherapy, heart–lung transplantation, DNase treatment and antibiotics. For those male sufferers who wish to father children intracytoplasmic injection is available. Despite available treatments CF is a lethal disease with patients rarely living beyond 40 years.

The *CFTR* gene product is a membrane protein that acts as an ion channel. The normal gene has been cloned and expressed in CFTR negative cells, restoring function. Animal models are available for the disease. Gene therapy for CF[206,207] focuses on treatment of the lungs, in spite of the fact that other organs may be affected, as it is the lung pathology that is difficult to control and will ultimately kill the patient. The lungs are easily accessible by use of aerosols such as those used in asthma treatment. Additionally, we know that with as few as 5% of cultured cells transduced, function is restored to the surrounding cells. This is because of an unexplained phenomenon, termed 'bystander effect', which is a function of cell communication. Finally, it is relatively easy to monitor the effectiveness of the therapy by simple skin tests.

Table 13.7 CF: a good candidate for gene therapy

- lungs are easily accessible
- gene is cloned
- animal experiments are possible
- cell culture results are promising
- communication between cells means that only 5% need to be infected to restore function = bystander effect
- easy tests for efficacy

13.3.1 Rationale for adenoviral vectors

Adenoviruses are natural infectious agents of the lungs; they are able to carry their genetic material into lung cells where their gene products are expressed. We have already looked at the properties of adenoviruses. The points to consider in this case are that we can make large numbers of replication-incompetent viruses for therapy but that the virus does not integrate and therefore any gene expression will be transient.

13.3.2 Early animal trials

Cell culture experiments had already shown that addition of adenoviruses containing the *CFTR* gene could restore function to CFTR negative cells. Animal models were then used to determine the effect of the gene therapy. A number of animals have been used, including mice, monkeys and cotton-tail rats. In essence, the results have shown that after a single treatment, gene expression could be detected for up to 6 weeks. The virus was, however, cleared fairly rapidly from the animal. No significant inflammation was seen, although the histology was not completely normal. Examination of faeces, urine, post-mortem tissue and air sampling suggested that no viral replication had occurred although some replication-incompetent virus was detected in faeces. Worryingly, when the animals were treated a second time they raised a strong immune response, rendering the therapy useless.

13.3.3 Is CF gene therapy 'safe but not useful'?

The animal studies suggested that the therapy was safe, but of little practical use. Nevertheless, trials then progressed to humans. CF trials with adenovirus in humans have all had problems with inflammation, transient systemic and pulmonary syndrome mediated by IL6 production and antibody response. This led to the following statement by one of the top researchers in this field: '*A lesson . . . is that despite extensive planning, animal studies and thorough review, pre-clinical studies DO NOT necessarily predict the response in humans to gene therapy vectors.*' Clearly they had not taken on board the results of the animal studies.

13.3.4 Problems also found in *in vivo* delivery

If we ignore for one moment the problems associated with the immune response, we can address the question of protein expression. Researchers have carried out a number of trials, three of which are shown in Table 13.8. One involved administration of virus to the nostril, one to part of the lung using a bronchoscope and the third used both techniques. All used patients over the age of 18 who were severely affected with little hope of other types of therapy. The total number treated was only 25. Trials 1 and 2 both used the same type of vector with the genes *E1* and *3* deleted whereas trial 3 used an adenovirus with only the *E1* gene deleted in an endeavour to lessen the problems associated with the immune response. Expression of the *CFTR* gene was not detected using protein analysis or skin tests. Using the very sensitive RT-PCR (reverse transcriptase PCR), which in theory will detect as little as a single message, 25% of nasal brushings and 33% of bronchial airway samples showed CFTR expression. The levels, however, were not therapeutic, but nevertheless these results encouraged researchers to look at different delivery systems to overcome the immunological problems seen with the use of viruses.

Table 13.8 Results of clinical trials for CF gene therapy

	USA 1	USA 2	USA 3
Numbers treated	10	12	3
Age	21+	18+	18+
Target tissue	left nostril and lung	part of lung	nostrils
Vector	E1+3 deleted	E1+3 deleted	E1 deleted
Success	none	none	some

A recent review of CF gene therapy[208] has stated that although clinical studies have demonstrated proof-of-principle for correction of the defect in CF patients using viral vectors, this has not yet resulted in gene therapy for patients. Problems associated with inefficient gene transfer and host immune responses caused by viral vectors have effectively halted trials. Non-viral delivery is being more thoroughly investigated using cationic liposome–plasmid DNA complexes and DNA nanoparticles to deliver the *CFTR* gene. These have shown some promise in phase I clinical trials but the levels of CFTR expression achieved in the respiratory epithelium were too low and only of limited duration. The authors suggest the need for improved strategies for efficient and prolonged expression of the introduced gene.

13.4 Animal trials with Factor IX[209]

Another area of great interest is the delivery of clotting Factors via gene therapy. We have already talked about the problems of isolating Factor VIII and related clotting factors from blood and the subsequent synthesis of recombinant proteins for treatment. However, there is still a potential risk of disease transmission even with recombinant factors and therefore many researchers are actively promoting gene therapy for the haemophilias. Currently these are being investigated in animal models.

An animal model is available for Factor IX deficiency, which causes haemophilia B. The *Factor IX* gene has been cloned into an adenoviral vector and introduced into mice by direct injection into a vein in the mouse tail. This was an interesting experiment in that delivery was not targeted to the organ which naturally produces the factor, the liver. Subsequent experiments showed that tail vein injection was as effective as direct injection into the organ. This is crucial for human therapy since injection into the liver would be unpleasant and potentially dangerous. Expression of the *Factor IX* gene was detected in liver for several weeks and tissue expression was also seen in lung, kidney and heart. The problems with the priming of the immune system were similar to those seen in the CFTR trials with antibody titres of 64 at the first injection and 4096 at the second. Subsequent therapy was ineffectual.

13.5 Adenoviruses have also been used to introduce genes into brain

Adenoviral vectors have been used to deliver genes to the brain, mainly in animal models of disease.[210] Original studies involved the direct injection of adenoviruses carrying the marker gene for β-galactosidase. After the experiment had concluded, post-mortem brain samples were harvested and stained for β-galactosidase activity. The distribution patterns showed that expression of the gene was limited to the site of injection and that a number of cells were killed by the procedure. In order to determine whether some areas of the brain accepted gene therapy better than others, adenoviral vectors containing marker genes were injected into rat brain. Expression was seen for a maximum of 3 months, the efficiency of transduction was 10% and the best results were seen when the gene was injected into the caudate nucleus.

Experimenters then moved on to transducing brain cells with adenoviral vectors containing the *tyrosine hydroxylase* gene, which is involved in the synthesis of L-dopa. In tyrosine-hydroxylase-negative rats treatment restored function for 2 months in 33% of the rats, which could be of significance to sufferers of Parkinson's disease where problems lie in the synthesis of L-dopa. However, there have been reports that adenoviruses cause inflammation of the brain, which is clearly a poor trade-off. In order to overcome problems of repeated injection, another group has described the production of mutant packaging cells that grow only at temperatures below 35 °C. These can be cultured indefinitely *in vitro* and transduced to carry the *TH* gene. On injection into the patient the cells continue to produce the TH but do not grow, as 37 °C is above their permissive growth temperature. The original animal studies suggest a 70% decrease in symptoms, but as yet human clinical trials have not followed.

13.6 Duchenne's muscular dystrophy

Duchenne's muscular dystrophy (DMD) is a disease that is untreatable and kills the young boys it affects. This muscle-wasting disease is caused by a defect in the dystrophin molecule which causes weakening of muscle strength. Additionally, victims lay down within their muscles collagen fibres that then cause immobility. It is X-linked and arises by virtue of many different mutations of the *dystrophin* gene. More than 25% of all those affected are the result of new mutations within the gene. Thus, preventive strategies are problematical as affected children may be born to those with no previous family history. Sadly, the disease may not be apparent until the children reach 4 or 5 years of age, and families appear in the literature with two or three affected children as the transmission of the disease was not realised until the family was completed.

The problems associated with conceiving a therapy for DMD are immense: the gene is one of the largest in the human genome and is therefore difficult to manipulate. In a manner similar to that already discussed for the production of recombinant Factor VIII, scientists have constructed the smallest gene fragment that would yield an active product in order to reduce the gene size. The tissue to be transduced is also a problem since muscle cells are numerous, difficult to access and divide infrequently. Following abortive attempts to inject the gene into muscle cells, current work is focusing on isolating myoblasts, transducing these and returning them to the affected individual.[211] Most of this work is being carried out in animal models. It is likely that, even if gene therapy becomes available for DMD, it will not be useful for those already showing symptoms as the muscle damage cannot easily be reversed. This has stimulated calls for all male neonates to be tested for DMD, so that when gene therapy is a reality they can be treated before symptoms arise.

13.7 Problems with adenoviruses

There had been problems associated with immune clearance of adenoviral vectors and some signs of brain inflammation, but adenoviruses were thought to be much safer than the integrating retroviruses. In September

1999, in the USA, an adenoviral gene therapy trial went tragically wrong and caused the death of an 18-year-old called Jesse Gelsinger.[212] The teenager suffered a mild form of ornithine transcarbamylase (OTC) deficiency controlled with a low-protein diet and drugs. In a random trial he was given the largest dose of adenoviral vector, and within 4 days he was dead. The death uncovered a number of unreported incidents in trials in the USA. Some doctors did not report deaths that they attributed to the disease rather than the therapy: deaths are not uncommon if we trial therapies on patients in end-stage disease and therefore we cannot be wholly certain whether the deaths arise from the disease or the therapy.

Jesse Gelsinger's death also resulted in a lawsuit, a government investigation, the delay of some other clinical trials and the creation of a new regulatory process for gene therapy trials in the US. There were also problems associated with informed consent. Had Jesse been given all the relevant information? Was he informed that several other patients had experienced serious side effects from the therapy, or that three monkeys had died of a severe liver inflammation and clotting disorder when treated in a similar way?[213] Why was he taken onto a trial when his disease was being well controlled by conventional therapy? However, it should be noted that following investigation of 400 trials involving 4000 patients no other deaths could be attributed to the administration of adenoviral vectors. As a consequence of his death, the trial was stopped, and early in 2000 the FDA began investigating 69 other gene therapy trials in the US: 28 trials were reviewed, 13 required ammendment.

13.8 The uses of adeno-associated viruses

13.8.1 Haemophilia B treatment with *Factor IX* gene augmentation[214]

Two articles in published in 1999 in *Nature Medicine*[215,216] illustrated *haemophilia* gene therapy trials carried out on dogs using AAV vectors. Both papers described Factor IX deficiency. Using AAV they demonstrated 17 months' expression of Factor IX at levels which would be therapeutic in humans. A series of intramuscular injections were used at a single time-

point and no adverse reactions were seen. Success was measured by immu-
nofluorescence and DNA analysis. Immune responses against Factor IX
were said to be absent or transient. In dogs this method gave 'partial
response' for 8 months, although the authors state that the dose was a
tenth that given to mice based on body mass.

A phase 1/2 dose-escalation clinical study extended this therapy to
humans with severe haemophilia B. An rAAV-2 vector containing human
Factor IX was infused into the hepatic arteries of seven patients. It was
shown that there was no acute or long-lasting toxicity with vector doses
up to 2×10^{12} vg/kg. The highest therapeutic levels of Factor IX were
achieved with the highest dose, and expression at therapeutic levels was
limited to 8 weeks. There was a gradual decline in Factor IX levels, accom-
panied by a transient asymptomatic elevation of liver transaminases: this
resolved without treatment. There was destruction of transduced hepatocytes by cell-mediated immunity that was targeted at antigens of the viral
capsid. Thus, although preliminary studies suggested that rAAV-2 vectors
could transduce human hepatocytes *in vivo*, studies are required on immu-
nomodulation if long-term expression is to be achieved.[217]

In a trial in the USA,[218] intramuscular injections of AAV vector into
skeletal muscle of humans with haemophilia B were shown to be safe, but
higher doses were required to reach therapeutic levels. The major problem
appeared to be the retention of Factor IX in extracellular spaces of the
muscle and the limited ability of the host muscle cells to synthesise high
expression of active factor. Researchers returned to their mouse models
and by using AAV variants were able to direct expression to the liver: this
could provide a strategy to improve the efficacy of gene-based therapies
for haemophilia B.

13.8.2 AAV therapy for DMD

A research team in the Netherlands has published a strategy for *DMD*
gene therapy based on AAV.[219] The vector had components of both high-
capacity adenoviral and AAV elements. These dual hcAd–AAV hybrid
vectors contained the gene for dystrophin which was expressed in rat car-
diomyocytes *in vitro*, and was able to restore dystrophin synthesis in the
muscle tissues *in vivo* in mouse models. The authors suggest that this
vector may be useful in the treatment of DMD in humans. However, the
problems associated with large-scale delivery are not addressed.

13.9 Liposome vector trials

Liposomes seemed to provide a viable alternative to adenoviral vectors for CF gene therapy with far less chance of immune stimulation. However, in clinical trials, in which liposomes were used to deliver *CFTR* into the nostril of CF sufferers, none of the 9 treated nasal brushings had detectable mRNA. Clearly the procedure was inefficient, but subsequent experiments went on to show that increasing the numbers of liposomes used increased efficiency of transfer but were also toxic.

Liposomes have been used to deliver genes *in utero* in lambs suffering from ductus arteriosus. This system was used to transfect the relevant area of the heart with liposomes containing plasmids that express the gene for an mRNA that will sequester fibronectin mRNA binding protein, the root cause of the problem.[220] Many forms of liposomes have been trialled, though it is difficult to find examples in which therapies have been successful.[221] We will revisit these in Chapter 15 on cancer gene therapy.

13.10 Trials with polymer matrix delivery

In this technique, plasmids incorporating *growth factor* genes have been immobilised in polymers delivered at the site of bone injury. The gene-encoded growth factors stimulated bone growth and the plasmid retained activity for 6 weeks.[222]

Points to consider

How do we test these new therapies?
How will we test *in utero* therapies in humans?
How valid are the results obtained?
Read about the case of Jesse Gelsinger, the first person known to have died from gene therapy. Whose fault was his death?
Given the problems of gene therapy and the deaths recorded with both retroviral and adenoviral vectors, should we stop trials?
View the material at: http://www.pbs.org/newshour/bb/health/july-dec99/gene_therapy_splash.htm
Does this affect your views on gene therapy?

Notes

197 Munyon, W. *et al.* Transfer of thymidine kinase to thymidine kinaseless L cells by infection with ultraviolet-irradiated herpes simplex virus. *J. Virol.*, 1971, **7**, 813–20

198 Palmiter, R.D. *et al.* Dramatic growth of mice that develop from eggs micro-injected with metallothionein-growth hormone fusion genes. *Nature*, 1982, **300**, 611–15

199 http://www4.od.nih.gov/oba/Rdna.htm

200 www.bloodjournal.org/cgi/content/full/101/7/2563

201 Booth, C. *et al.* Management options for adenosine deaminase deficiency; proceedings of the EBMT satellite workshop (Hamburg, March 2006). *Clin. Immunol.*, 2007, **123**, 139–47

202 Aiuti, A. Gene therapy for adenosine-deaminase-deficient severe combined immunodeficiency. *Best Pract. Res. Clin. Haematol.*, 2004, **17**(3), 505–16

203 www.webconferences.com/nihoba/RAC%203-15-05_Candotti2.pdf

204 www4.od.nih.gov/oba/rac/SSMar05/pdf/FINAL%20X-SCID%20summary.pdf

205 http://gut.bmjjournals.com/cgi/content/full/46/1/136

206 Griesenbach, U. *et al.* Gene therapy progress and prospects: cystic fibrosis. *Gene Ther.*, 2006, **13**, 1061–7

207 Anson, D.S. *et al.* Gene therapy for cystic fibrosis airway disease – is clinical success imminent? *Curr. Gene Ther.*, 2006, **6**, 161–79

208 Rosenecker, J. *et al.* Gene therapy for cystic fibrosis lung disease: current status and future perspectives. *Curr. Opin. Mol. Ther.*, 2006, **8**, 439–45

209 Smith, T.A. *et al.* Adenovirus mediated expression of therapeutic plasma levels of human factor IX in mice. *Nat. Genet.*, 1993, **5**, 397–402

210 www.bentham.org/cgt/samples/cgt5-1/0006Q.pdf

211 van Deutekom, J.C. and van Ommen, G.J. Advances in Duchenne muscular dystrophy gene therapy. *Nat. Rev. Genet.*, 2003, **4**, 774–83

212 www.cmaj.ca/cgi/content/full/164/11/1612

213 http://archives.cnn.com/2000/HEALTH/01/22/gene.therapy/

214 www.mssm.edu/msjournal/71/71_5_pages_305_313.pdf

215 Herzog, R.W. *et al.* Long-term correction of canine hemophilia B by gene transfer of blood coagulation factor IX mediated by adeno-associated viral vector. *Nat. Med.*, 1999, **5**, 56–63

216 Snyder, R.O. *et al.* Correction of hemophilia B in canine and murine models using recombinant adeno-associated viral vectors. *Nat. Med.*, 1999, **5**, 64–70

217 Manno, C.S. *et al.* Successful transduction of liver in hemophilia by AAV-Factor IX and limitations imposed by the host immune response. *Nat. Med.*, 2006, **12**, 342–7

218 Montier, T. *et al.* Non-viral vectors in cystic fibrosis gene therapy: progress and challenges. *Trends Biotechnol.*, 2004, **22**, 586–92

219 Schuettrumpf, J. *et al.* Factor IX variants improve gene therapy efficacy for hemophilia B. *Blood*, 2005, **105**, 2316–23

220 Mason, C.A. *et al.* Gene transfer in utero biologically engineers a patent ductus arteriosus in lambs by arresting fibronectin-dependent neointimal formation. *Nat. Med.*, 1999, **5**, 176–82

221 Hart, S.L. Lipid carriers for gene therapy. *Curr. Drug Deliv.*, 2005, **2**, 423–8

222 Bonadio, J. *et al.* Localized, direct plasmid gene delivery in vivo: prolonged therapy results in reproducible tissue regeneration. *Nat. Med.*, 1999, **5**, 753–9

14

Gene silencing technologies

These are four different techniques involved in gene silencing or 'turning off' genes. They work on different molecules at different stages of the cell cycle. These techniques are designed as nucleic acid drugs and would be used like other forms of drug therapy. There is usually no long-term effect and the molecules would have to be administered at intervals to the patient to achieve the required effect. There have been papers suggesting a virus-based delivery to give long-term expression, but this has as many problems as viral vectors in gene therapy.

Originally, this was a costly procedure, but the techniques now compare well with other therapies in terms of cost. The methods are applicable to the treatment of cancers, infections and, maybe, in the future, to dominant diseases. They have also proven to be useful tools, allowing us to turn off genes in animal models and tissue culture and to determine the outcome of lack of expression. We will address each type of molecule separately as the mechanisms of action are not identical.

14.1 Antisense therapy

Antisense molecules are short single-stranded pieces of DNA which are complementary to the mRNA that we wish to silence. The antisense molecule essentially blocks translation of the mRNA into a functional protein. Antisense molecules were originally discovered in bacteria

Molecular Therapeutics: 21st-century Medicine by Pamela Greenwell and Michelle McCulley.
© 2007 John Wiley & Sons, Ltd

and viruses where they play a natural role in control of gene expression. Simplistically, the organism transcribes from both strands of the DNA of some genes, producing a functional mRNA and a complementary antisense molecule. If synthesised in equal amounts there is no gene activity. If more of the mRNA is produced than the antisense, then protein production will occur. Clearly there are sophisticated controls which regulate the transcription of each molecule, and which in turn are affected by the environment. This type of control is universal in bacteria and viruses.

In mice and humans, the *insulin-like growth factor 2 receptor (Igf2r)* gene that is inherited from the father produces an antisense RNA that blocks synthesis of the Igf2r mRNA.[223] Imprinting results in the inactivation of the genes from either the mother or the father (dependent on the gene). Igf2r is an example of an imprinted gene and is one of a number of examples of antisense transcripts that have been identified in several imprinted genes, suggesting that they may play an important role in the process of genomic imprinting.

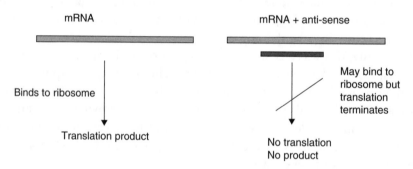

Figure 14.1   Mechanism of antisense action

Antisense molecules are oligonucleotides, complementary to a specific mRNA to which they bind, preventing translation. In fact, the oligonucleotide binds to the mRNA, the ribosome reads mRNA until it reaches the oligonucleotide and the ribosome and RNA fall apart. This is important to remember since we will not completely block transcription by use of 10–20 bases antisense molecules that target sequences within the message. If an incomplete message is translated it is important that the resultant protein fragment is not active. It is possible to circumvent this problem by designing molecules that are antisense to sequences at the beginning (DNA with free 3' hydroxyl group) of the mRNA. It has also become clear that RNase H present in the host cells recognises the mRNA-oligonucleotide hybrid and degrades the message. Therapeutically, this adds to the efficacy of the technique.

14.1.1 Modifications of antisense molecules

If we inject unmodified oligonucleotides into a human they would be destroyed almost instantly by the host nucleases. Therefore, for use in therapy the oligonucleotides must be altered to ensure that they are not degraded by nucleases. Figure 14.2 is a cartoon of DNA structure; you will note that some of the oxygens of the phosphate group are used in the formation of the linkage between the sub-units, whereas the bridge and others are not.

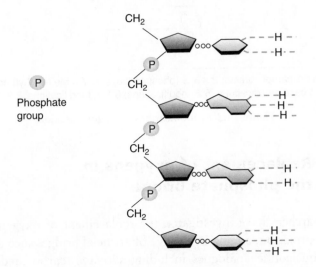

Figure 14.2   Cartoon of oligonucleotide

Modifications may be made to the molecule provided that these do not affect the binding capacity of the molecule to its target.[224] Modifications have been used to prevent degradations which involve changes in those oxygens of the phosphate which are not involved in bridge formation. These are shown in Table 14.1.

Table 14.1   Modified oligonucleotides

- methylphosphonate oligonucleotides, called MATAGENES (masking tape for gene expression)
- phosphorothioates/phosphoradithioates
- phosphoamidates
- phosphate esters, alkyl derivatives

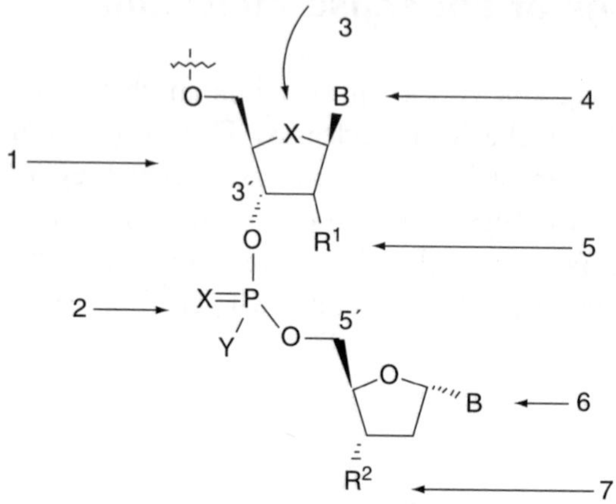

Modifications: 1 carbohydrates; 2 X = S (phosphothio) and Y = Me (methyl) or alkyl;
3 4' S,C,N; 4 base modifications; 5 2' modifications; 6 1' α modifications; 7 3' modifications

Figure 14.3   Modifications to oligonucleotides

14.1.2 Replacement of oxygens in the phosphate bridge

Some researchers have investigated the replacement of oxygens involved in bridge formation, adding in a range of artificial bridges such as: dephospho internucleotide analogues including siloxane, carbonated, carboxymethyl ester, acetamide, carbamate and thioester and plastic bridges made using polymers such as polyvinyls.

14.1.3 Modifications can be made to the bases themselves

These usually involve changing the anomeric configuration of the nucleosides from β to α or by incorporation of the novel deoxyinosine molecule.

Both induce changes that render the oligonucleotide more resistant to attack by nucleases.

14.1.4 Other types of modifications

Other types of modifications have involved the addition of prosthetic groups to the oligonucleotide. For example, if Fe-EDTA (iron linked to the chelating agent EDTA) is complexed to the oligonucleotide, a catalytic antisense molecule is generated which binds to the mRNA and then cleaves it, releasing the antisense oligonucleotide, allowing it to bind to another mRNA molecule. Other additions include acridine dye which increases oligonucleotide uptake: acridine is, however, a mutagen!

14.1.5 The ideal oligonucleotide

The characteristics of the ideal oligonucleotide are listed below. Unlike our ideal vector for gene augmentation therapy, the ideal oligonucleotide can be made successfully for use *in vivo*.

Table 14.2   Characteristics of the ideal oligonucleotide

1. should give a stable complex with RNA/ DNA
2. should show specific targeting
3. must have a long half-life, resistant to nuclease
4. must easily pass through cell membranes to reach the nucleic acid
5. must not accumulate in organs and organelles
6. should be about 18–20 bp

14.1.6 Uptake

The uptake of the molecules is rapid, specific and targeted to the cells and nuclei. Resistance to nucleases of modified oligonucleotides results in a high concentration of oligonucleotides within the nucleus remaining for at least 3 hours. Receptor-mediated binding can be engineered by delivering oligonucleotides attached to growth factors and, if necessary, long-term expression can be achieved by introducing into the cells viruses which express the oligonucleotide as mRNA. In this case we would have the same problems as those described in the section on gene therapy. It is clear that such techniques would be required if we wished to have long-term expression for the treatment of dominant diseases.

However, simple delivery into the tail vein of a mouse resulted in therapeutic levels of oligonucleotides being detected in all organs with the exception of the brain (due to the blood–brain barrier) within 30 minutes. In mice, antisense molecules have been emulsified in creams and rubbed into the ears with good results. Liposomes have also been used to deliver oligonucleotides, but in this case there is preferential uptake by the spleen and liver. Experiments have shown that:

- mRNA is made in the nucleus and transported to cytoplasm to synthesise proteins on ribosomes

- oligonucleotides must pass through membrane by endocytosis into the cell

- in cell culture, 7–10% of oligonucleotides are inside cells and within nucleus in 15 minutes after addition to the medium

- only 6% of oligonucleotides taken up by the cells are broken down by nucleases

- oligonucleotides remained in nucleus for at least 3 hours

- oligonucleotides can be delivered within liposomes or attached to factors to allow receptor-mediated binding

- viruses can be made which synthesise oligonucleotides continuously; if they are used to infect cells, we get long-term expression

- in mice, injected oligonucleotides were found in all organs and cells with the exception of the brain at therapeutic levels within 30 minutes of injection

- in mice DNA in cream applied to ears gave successful delivery

- liposomes have been used to deliver oligonucleotides to give cellular concentrations up to 20 nM (nanomolar) *but* preferential uptake is by spleen and liver

- immunoliposomes using an antibody for 'homing' have also been used

- polylysine has been suggested as a carrier with high affinity for cell membranes *but* is toxic.

14.1.7 Uses of antisense oligonucleotides

The uses of antisense therapy revolve around the treatment of acquired rather than inherited diseases. In many cases, single treatments are used either to kill the infective organism or to remove cancer cells. The problem of treating, for example a dominant disorder, using this approach is that, since mRNA is continually produced, often at high level, vast amounts of oligonucleotides would be required to turn off every affected message. Since we are as yet unsure as to the long-term effects of the treatment, antisense may not provide a sensible approach. If we could deliver viral vectors efficiently to cells, these could be manipulated to produce antisense molecules, offering hope of long-term gene correction for sufferers of dominant disorders.

Table 14.3 Potential applications of antisense therapy

- anti-viral therapy
- anti-bacterial therapy
- anti-parasitic therapy
- purging of cancer cells *ex vivo*
- treatment of skin disease
- inhibition of inflammation
- suppression of oncogenes
- suppression of dominant genes
- identification of targets for drug or gene therapy

14.1.8 Examples

One of the first experiments carried out involved the use of antisense molecules complementary to part of the 35-nucleotide (nt) sequence present at the 5′ end of all trypanosome mRNAs (mini-exon sequence) to kill *Trypanosoma brucei* in the blood stream of cattle. The oligonucleotides were linked to acridine dyes which aided the uptake of the oligonucle-otide.[225] The experiment was successful and heralded the use of antisense technology in microbial therapeutics.

In humans most of the interest is focused on HIV and, as we will see, many researchers have tried to target virus-specific genes in order to prevent viral replication or transmission; for example scientists targeted antisense to *gag* mRNA which is crucial for viral replication. The life of the molecule was 5 days. It did not kill the organism but shut down repli-cation with no side effects.

14.1.8.1 Ear creams have been used to deliver oligonucleotides in mice

As discussed, delivery of the oligonucleotides is relatively simple. Experiments in mice, involving rubbing emulsions containing antisense molecules (in this case methyl-phosphonate oligonucleotides), to herpes simplex virus (HSV) have resulted in proven absorption of the DNA and both prevention and treatment of HSV lesions. Clearly this type of therapy would be useful in treating HSV infections in humans and would be easy to produce and administer. The nature of the oligonucleotides makes them heat-stable and therefore it is easy to envisage such therapies being sold in much the same way as other topical treatments.

14.1.8.2 Malaria

Since successes were achieved with trypanosome infections researchers have also investigated the possibility of using antisense oligonucleotides to treat malaria. Currently, malaria is becoming more difficult to treat because of the emergence of resistant strains. The protocol has used the enzyme dihydrofolate reductase–thymidylate synthase (DHFR–TS),[226] which is involved in pyrimidine production. Without this enzyme the parasite would be rendered replication-incompetent. In the past researchers have tried to block this enzyme using traditional drugs; however, humans also possess a DHFR-TS crucial to survival which is affected by traditional enzyme inhibitors. The enzymes produced by the parasite and by man are very similar – too similar at the protein level to allow specific targeting by drugs. At the gene level there are significant differences in the sequence. These can be targeted by antisense techniques, specifically turning off the malarial enzyme whilst leaving the host enzyme intact.

14.1.8.3 Antisense also used to turn off genes in vivo

In rabbit models of heart disease, scientists have investigated the problems associated with blood vessel occlusion following angioplasty. Angioplast devices are used to restore blood flow in cases of vessel occlusion, but in some patients the device itself then leads to further occlusion and in some cases death. The problem appears to involve 'irritation' of the endothelial cells of the vessel, which then produce the proto-oncogenic protein c-myc which in turn stimulates cells to proliferate. In the rabbit model, once the angioplast device was in place, antisense oligonucleotides that turn off expression of the protein were delivered. Following treatment there was a substantial reduction in the numbers of rabbits that went on to develop device-related occlusions.[227] Studies in humans have also started.[228]

14.1.8.4 Antisense therapies in humans

Every day, more antisense therapies are described and Table 14.4 gives a list of some of those that have reached the human clinical trials stage or beyond. We have already mentioned treatment of HIV and malaria. There are also treatments available for cytomegalovirus (CMV) and human papilloma virus. Indeed, a commercial company is marketing antisense CMV for the treatment of retinitis seen in HIV patients. Researchers are looking at antisense molecules to reduce clotting in thrombotic patients and to reduce c-myc expression in cancer cells. A measles antisense is in trials to attempt to clear persistent infections from patients where this remains a problem. This is important, as the results of viral re-activation can be catastrophic.

Table 14.4   Antisense molecules in trials

- antisense to HIV gag mRNA and revRNA
- anti-HPV
- measles virus antisense sequences used specifically to cure cells persistently infected with measles virus; currently in trials
- antisense used for treatment of CMV retinitis phase III study of topical agent to prevent transcription *in vivo*
- anti-malarial DHFR-TS, much more specific than drugs which target enzyme – gene is different enough to enable oligonucleotides to be specifically designed
- anti-coagulant to prevent clotting
- c-myc resistance to cancer drugs is associated with high levels of myc expression – therefore turn off myc and make cells more susceptible to drugs such as cisplatin and vinblastine; works in culture

In May 1999, *Nature Biotechnology* published an article on antisense therapeutics (**17**, 403) Table 14.5 lists those companies who are still working in the area.

At the beginning of this work there was a real problem with supply of enough oligonucleotides to make treatment economical. New methodologies have made such treatments more feasible. Nevertheless, since mRNA may have thousands of copies produced per cell per hour the amount of oligonucleotide needed to treat each patient is clearly an important consideration. Much of the work that has been successful has involved microorganisms, the genomes are relatively easy to target and elimination of infection can be easily attained.[229,230]

Table 14.5   Selected companies with antisense programmes

Company	Programme
AVI Biopharma	restenosis inhibitor about to enter phase 3 trials
	hepatitis C in late phase 2
CV Therapeutics	restenosis and vascular therapies
Cytoclonal Pharmaceutics	localisation of optimal binding sites
Genta	cancer
Gilead Sciences	code blocker technology
Human Genome Sciences	drugs from genomics
Hybridon	AIDS, hepatitis B, hepatitis C, HPV, cancer
Isis Pharmaceuticals	ulcerative colitis
Lynx Therapeutics	chronic myeloid leukaemia
Pangea	resources
Rhone Poulenc	cancer

Table 14.6   Antisense therapies in phase 3 and beyond

Drug	Status	Basis of action	Disease indication
Vitravene – Isis Pharmaceuticals	FDA-approved	Inhibitor of immediate early region 2 (IE2) of human cytomegalovirus	Cytomegalovirus retinitis in AIDS patients
Affinitak – Isis Pharmaceuticals	Phase 3 in combination with carboplatin and paclitaxel	Inhibitor of protein kinasec-alpha (PKC-alpha) expression	Stage IIIb or Stage IV non-small cell lung cancer
Alicaforsen – Isis Pharmaceuticals	Phase 3	Inhibitor of intracellular adhesion molecule-1 (ICAM-1)	Crohn's disease

Table 14.6   Continued

Drug	Status	Basis of action	Disease indication
Macugen – Eyetech and Pfizer	Phase 3 +/– photodynamic therapy	Inhibitor of vascular endothelial growth factor (VEGF)	Age-related macular degeneration
Genasense – Aventis and Genta	Late-stage phase 3 in combination with dexamethasone	Inhibitor of B-cell leukaemia/ lymphoma 2 (Bcl-2) protein	Malignant melanoma
Genasense – Aventis and Genta	Phase 3 in combination with fludarabine and cyclophosphamide	Inhibitor of B-cell leukaemia/ lymphoma 2 (Bcl-2) protein	Chronic lymphocytic leukemia
Genasense – Aventis and Genta	Phase 3 + dacarbazine	Inhibitor of B-cell leukaemia/ lymphoma 2 protein	Multiple myeloma

14.1.9 Catalytic antisense molecules

As we have discussed, catalytic antisense molecules may offer us a way around the problem of the large amounts of oligonucleotides required and the possibility of only partially silencing genes. Fe-EDTA bound to the oligonucleotide leads to destruction of the mRNA and theoretically the oligonucleotide will be freed for further interactions. The consequences of such strategies have not been thoroughly investigated.

14.2 Triple helix (triplex) technology

In order to overcome the problems associated with antisense oligonucle-otides researchers have turned their attention to intervention at an earlier stage, preventing transcription rather than translation. In this case, a mol-ecule that could bind to DNA in the chromosome could be used to turn off the gene. Theoretically one oligonucleotide molecule would be required per cell treated. At this stage, you should have realised that there is a problem here. Chromosomes are double-stranded and traditionally we talk about single strands binding together to form stable complexes, so how does this therapy work?

Essentially it has been shown that an oligonucleotide binds to double-stranded DNA by Höögsteen binding, T binds to an A–T pair and G or C binds to a G–C pair and inactivates the DNA by preventing transcription. The resultant triple helix cannot be transcribed and hence the genes cannot be expressed.[231]

3'-TTTCTTCTTCT   Watson–Crick H-bonding pyrimidine strand
 AAAGAAGAAGA-5'   purine strand
 TTTCTTCTTCT-5'   Höögsteen H-bonding pyrimidine strand

Figure 14.4   Höögsteen binding

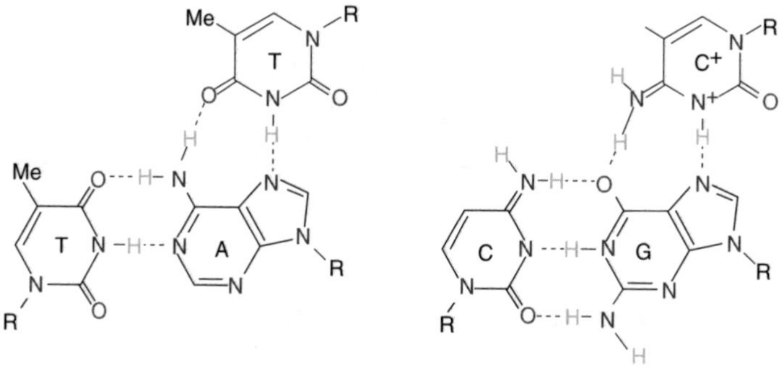

Figure 14.5   Formation of triplex

Figure 14.6   Molecular model of a triple helix[232]

This technique appears to be very effective when it works, as there is only one DNA target per cell. Originally, this was seen to be suitable for treatment of dominant diseases since low levels of oligonucleotides would be required and the effect would persist until the cell divided. For even longer-term expression, for example in the treatment of stem cells and their progeny, researchers suggested that viruses could be used to carry DNA-encoding oligonucleotide into the cell. Provided the virus integrated, the oligonucleotide encoded would be continually expressed. Of course, we must remember that we cannot yet achieve successful gene delivery of this type.

There have been claims that triplex formation could be used for directed genome modification, with the ultimate goal of repairing genetic defects in human cells. Several studies have demonstrated that treatment in mammalian cells can bring about DNA repair and recombination in such a way that desired sequence changes could be introduced.[233]

14.2.1 Problems

Clearly, since all dominant diseases have not been cured, there must be a problem with this type of therapy. To achieve therapy in human cells *in vivo*, an oligonucleotide must enter the nucleus and bind to the DNA which is supercoiled and has histones associated with it. This is in fact difficult to conceive. The only successful experiments have involved triplexes that bind to promoter sites that are more exposed prior to transcription than the genes themselves.

Therefore, most of the work published has in fact been used to target microbes. The microbial and viral genomes are far less complex than the human and therefore such therapy is much more successful.

14.2.2 Advantages over antisense strategies

We have talked about the fact that fewer molecules would be needed to achieve an effect with triplexes rather than antisense and that this could result in a long-acting low-dose therapy. Another potential advantage, providing we are dealing with microbes, is that the addition of an N-bromoacetyl electrophile attached to the 5'-phosphate group of a purine-rich oligonucleotide mediates binding of the oligonucleotide to the double-stranded DNA sequence, resulting in specific alkylation[234] that turns off genes in the vicinity of binding. In the treatment of microbial infections this would be an advantage; in human therapy it could result in nearby genes being accidentally inactivated with potentially dire consequences.

14.2.3 Experimental data

14.2.3.1 Insulin-like growth factor

In cultured cells,[235] oligonucleotides have been used to target binding to the 23 bp upstream of the promoter of the *insulin-like growth factor* gene. When the triplex forms it prevents RNA polymerase binding, resulting in arrested transcription. Although problems were noted with poor uptake of the oligonucleotide into the nucleus, the majority being held in endosomes and lysosomes, the treatment prevented transcription of the gene. Tumour cells in culture when re-implanted *in vivo* lost tumorigenicity and elicited tumour-specific immunity, leading to elimination of established tumours.

Targeting simian and human immunodeficiency viruses

Here, researchers are targeting the integrated viral DNA. The targets are either the *vpx* gene or the *PPT (polypurine tract)* gene. These are essential for viral replication. However, there have been no reports of spectacular successes.

14.2.3.2 Sequence-selective recognition and cleavage of double-helical DNA

This is the same type of strategy described for antisense molecules where Fe-EDTA linked to the oligonucleotide mediates cleavage of the DNA by induction of an artificial nuclease. Again, this could be useful in the treatment of acquired diseases but would be disastrous if attempted on the host DNA.

Table 14.7 shows the numbers of US patents lodged by companies that involve triplex technology.[236]

Table 14.7 US patents involving triplex agents

Patent	Date	Company	Project
US7070933	2006	Gen-Probe Incorporated	Inversion probes
US7057027	2006	ISIS Pharmaceuticals, Inc.	Enhanced triple-helix and double-helix formation
US6962783	2005	Isis Pharmaceuticals, Inc.	Enhanced triple-helix and double-helix formation
US6951931	2005	ISIS Pharmaceuticals, Inc.	Pyrimidine derivatives
US6875593	2005	ISIS Pharmaceuticals, Inc.	Enhanced triple-helix and double-helix formation
US6800743	2004	ISIS Pharmaceuticals, Inc.	Pyrimidine derivatives
US6683166	2004	ISIS Pharmaceuticals, Inc.	Modified internucleoside linkages
US6670393	2003	Promega Biosciences, Inc.	Carbamate-based cationic lipids
US6489464	2002	Hybridon, Inc.	Branched oligonucleotides as pathogen-inhibitory agents
US6414127	2002	ISIS Pharmaceuticals, Inc.	Pyrimidine derivatives
US6410702	2002	ISIS Pharmaceuticals, Inc.	Modified internucleoside linkages (II)
US6380368	2002	ISIS Pharmaceuticals, Inc.	Enhanced triple-helix and double-helix formation
US6372427	2002	Hybridon, Inc.	Cooperative oligonucleotides
US6133024	2000	Aventis Pharma S.A.	Gene expression control
US6028183	2000	Gilead Sciences, Inc.	Pyrimidine derivatives and oligonucleotides
US6007992	1999	Gilead Sciences, Inc.	Pyrimidine derivatives for labelled binding partners
US5948634	1999	The General Hospital Corp.	Neural thread protein gene expression and detection of Alzheimer's disease
US5948888	1999	The General Hospital Corp.	Neural thread protein gene expression and detection of Alzheimer's disease
US5872105	1999	Research Corporation Technologies Inc.	Single-stranded circular oligonucleotides useful for drug delivery
US5830653	1998	Gilead Sciences, Inc.	Methods of using oligomers containing modified pyrimidines

Table 14.7   Continued

Patent	Date	Company	Project
US5830670	1998	The Gen. Hospital Corporation	Neural thread protein gene expression and detection of Alzheimer's disease
US5817781	1998	Gilead Sciences, Inc.	Modified internucleoside linkages (II)
US5800984	1998	Idexx Laboratories, Inc.	Nucleic acid sequence detection by triple-helix formation at primer site in amplification reactions
US5739308	1998	Hybridon, Inc.	Integrated oligonucleotides
US5683874	1997	Research Corp. Technologies	Single-stranded circular oligonucleotides capable of forming a triplex with a target sequence
US5670634	1997	The General Hospital Corp.	Reversal of β/A4 amyloid-peptide-induced morphological changes in neuronal cells by antisense oligonucleotides
US5646261	1997	Hoechst Aktiengesellschaft	3'-derivatised oligonucleotide analogues with non-nucleotidic groupings, their preparation and use
US5616461	1997	Dana-Farber Cancer Institute	Assay for antiviral activity using complex of herpesvirus origin of replication and cellular protein

14.3 Ribozymes

Ribozymes are naturally occurring RNA sequences that have specific functional properties and can cleave RNA targets to which they hybridise. It is known that[237] *in vitro* a number of small RNAs (viroids, satellite RNAs and virusoids) pathogenic to plants, an RNA transcript from *Neurospora* mitochondrial DNA, and an animal virus, hepatitis-D (HDV), undergo a self-cleavage reaction in the absence of protein. The reaction requires magnesium and produces 2′3′-cyclic phosphate and 5′OH termini. The mechanism involves several different catalytic motifs: the hammerhead, hairpin, and axehead or pseudoknot. These molecules cleave at GUC in the target sequence. Since GUC is present in every mRNA these can be used as universal cutting tools. The most important feature therefore is to target the ribozyme to the appropriate mRNA. This can be difficult since even mRNA has a structure with areas inaccessible to ribozymes.

14.3.1 Examples of ribozyme therapies[238]

14.3.1.1 Human papilloma virus (HPV)

Ribozymes have been used to target HPV and early experiments looked at three potential ribozymes. All showed maximal binding activity to their targets within an hour and the ribozymes cleaved HPV RNA *in vitro* in infected cells. This work has not yet moved into clinical trials. In the main this is due to problems associated with production of nuclease-resistant ribozymes at concentrations likely to have effect *in vivo* and the development of vaccines targeted at HPV prevention.

14.3.1.2 Ribozymes used to cleave Bcr-Abl oncogene construct

Bcr-Abl is formed by the translocation of the tips of the long arms of chromosomes 9 and 22 found in chronic myeloid leukaemia (CML). The translocation generates a novel transcript which produces a protein kinase, P210. This acts to phosphorylate and hence activate other molecules inappropriately. *In vitro*, ribozyme treatment leads to loss of protein kinase activity.

A ribozyme which targeted the junction sequence of bcr-abl was shown, in animal models and cell culture, to specifically cleave the bcr-abl mRNA. This in turn induced apoptosis in CML cells. A retroviral system was used to deliver a modified ribozyme, a maxizyme, to cells.[239] Modified cells expressing the vector were injected into tail veins of immune-deficient mouse models of leukaemia. Animals treated with the control vector died between 6 and 13 weeks afterwards due to diffuse leukemia whereas the 8 animals treated with the *bcr-abl* ribozyme showed no evidence of leukemia 8 weeks after inoculation. It has been suggested that this type of strategy could be useful for purging bone marrow in cases of CML treated by autologous transplantation. A similar approach has been used to target the p190 fusion protein seen in ALL (acute lymphoblastic leukaemia) patients.[240] This work is still in its infancy and may only ever be applicable to *in vitro* work.

14.3.1.3 HIV therapies

There have been plenty of trials of ribozymes for HIV. In an elegant piece of work researchers infected cells with murine retrovirus containing a sequence for an anti-HIV-1 ribozyme. When the infected cells were challenged with HIV 1 they were resistant whereas no resistance to infection was seen with HIV 2.[241] Trials have not yet taken place. Who would be willing to undergo such a 'vaccination' programme given the current problems associated with viral vectors?

Other workers have tried ex *vivo* retroviral transduction, to introduce an HIV-1 ribozyme (RRz2) targeting the HIV *tat* gene and a control construct (LNL6) into CD34+ haematopoietic stem cells (HSC). Transduced autologous CD34+ cells were infused into 10 patients in a phase I study. After 2.5 years, the ribozyme gene sequence and expression were detected by a sensitive polymerase chain reaction (PCR) assay. The effect was proportional to the dose of infused stem cells in a transduced-CD34+ cell dose-dependent manner.[242] Again, this is demonstration of the concept, but as yet these therapies are a long way from being used in patients.

14.4 Small interfering RNAs (siRNAs)[243]

Until recently, only single-stranded RNA was thought to have an effect on post-transcriptional gene expression. This single-stranded material was highly susceptible to degradation by RNase. The RNA interference silencing (siRNA) phenomenon, which involves double-stranded RNA, was first observed in plants, and in 1998 the critical role of double-stranded RNA (dsRNA) was recognised in the nematode worm *Chaenorhabditis elegans*. The original observations involved the breeding of petunias.[244] These plants exist in a range of different colours and breeders were trying to make dark purple or black flowers. The scientists introduced the gene for purple pigment into pale petunias under the control of a strong promoter. Instead of the expected dark purple plants the offspring were white. This suggested that they were seeing quelling, a phenomenon seen in fungi which causes co-suppression of genes. The same mechanism has been found in a number of plants and animals and has *in vivo* roles in viral defence and transposon silencing. Basically, the mechanism involves the introduction of double-stranded RNA as a tool to silence gene expression.

The double-stranded RNA molecules were termed 'small interfering RNAs' or siRNA. In experiments using *C. elegans,* use of an siRNA in the gut caused gene silencing throughout the nematode. In this case the targets were mRNAs encoding splicing factors and mRNA export proteins. Interestingly, with these, though not other siRNAs, the effects were lethal to offspring.[245] In mice, scientists could turn off 81% of introduced *luciferase* genes with siRNA and in cell culture siRNA has shown suppression of the HIV life cycle.[246]

In a separate experiment to view the effects of antisense molecules, researchers used two controls, the sense molecule and a duplex of sense and anti sense oligonucleotides. We would expect the antisense to turn off gene expression, and the sense, which is identical to the RNA, should have no effect. A duplex of sense and antisense should form a double-stranded molecule which would not bind to the target RNA. Although antisense worked well, the duplex of sense and antisense was even more efficient, requiring far fewer molecules to achieve silencing. This was a surprising and reproducible effect. Clearly, if we understood how this worked we could harness the method to turn off genes. It was originally thought that the introduced gene caused methylation and gene silencing. This mechanism could be triggered by certain viruses and once triggered it was mediated through small diffusible trans-acting molecules (double-stranded RNAs). Experiments showed that there is message made but that it is rapidly degraded.

The property of dsRNA to induce RNA interference (RNAi) has become a powerful tool for scientists to study the function of genes in many lower organisms. However, it was originally difficult to apply this technology to mammalian cells as dsRNA provokes a non-sequence-specific response via the interferon pathway that is normally used by mammalian cells to combat a viral infection. This response causes cells to undergo apoptosis to avoid spreading the perceived virus to their neighbours. New strategies have been developed for triggering RNAi, to circumvent this limitation in applying RNAi to mammals.

In the RNAi pathway, long dsRNA is cleaved into smaller segments of 21 to 23 nucleotides, called small interfering RNA (siRNA), by type

III RNases that act as guides, binding to the corresponding segments of an mRNA. Following binding to the mRNA, the siRNA guide recruits cellular enzymes that cleave and destroy the mRNA. Recently it has been shown that siRNAs associate with cellular proteins to form a protein-based complex called RISC (RNA-induced silencing complex). Within the RISC complex, the two strands of the siRNA become separated, so that they can target complementary sequences in mRNAs involved in a disease. After pairing with an siRNA strand, the targeted mRNA is cleaved and undergoes degradation, thereby interrupting the synthesis of the disease-causing protein. The RISC complex is naturally stable within the cell, enabling siRNAs to cut multiple mRNA molecules consecutively and, therefore, suppressing protein synthesis in a potent and targeted way.

Table 14.8   Mechanism of siRNA action[247]

1. Small interfering RNA (siRNA), a 21–25-base-pair RNA strand, is targeted to a specific gene.
2. Within cells, siRNA unwinds and is incorporated into RISC, a stable protein–RNA complex.
3. siRNA is directed to a targeted messenger RNA (mRNA) that is known to be involved in a disease pathway.
4. The mRNA undergoes degradation, thereby interrupting the protein synthesis of the targeted gene.

In May 2001, Elbashir *et al.*[248] reported in *Nature* that siRNAs could be used to specifically suppress expression of genes in mammalian cells, with no unwanted non-specific damage to the cell. The small size of the siRNAs enables them to suppress gene expression without provoking undesirable cellular responses. siRNA-based drugs can distinguish between a normal and a mutated sequence that differ by only one nucleotide.

Since its development, the RNAi technique has revolutionised studies of non-mammalian systems. The RNAi method enables a researcher to down-regulate expression of a targeted gene without side effects, thereby revealing the function of the silenced gene.

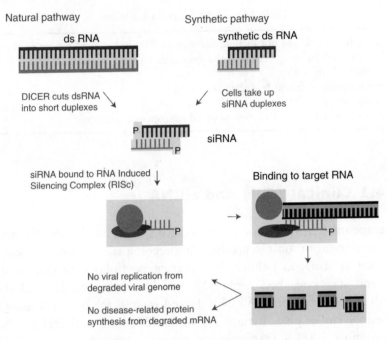

Natural pathway   Synthetic pathway

ds RNA   synthetic ds RNA

DICER cuts dsRNA   Cells take up
into short duplexes   siRNA duplexes

siRNA

siRNA bound to RNA Induced   Binding to target RNA
Silencing Complex (RISc)

No viral replication from
degraded viral genome

No disease-related protein
synthesis from degraded mRNA

Figure 14.7   Mechanism of siRNA action

Although siRNAs were able to elicit strong and specific suppression of gene expression in different mammalian cell lines, this effect is only transient, because of the short half-life of synthetic RNAs, which severely limits their applications. To overcome this limitation, companies have designed efficient and simple-to-use vectors, called psiRNAs, that allow the production of siRNAs within the cells. psiRNAs are used to insert a DNA fragment of approximately 50 bases, designed in such a way that after transcription from an RNA polymerase III promoter it will generate siRNAs with a hairpin structure. Short hairpin RNAs are more stable, and since siRNAs are continuously expressed within the cells, this method permits long-lasting silencing of genes of interest.

However, before siRNA can be used as an efficient and safe drug in humans, a number of issues will have to be resolved. In particular, proper methods of delivery must be developed. Delivery is the most important problem to have beset gene augmentation therapy. How will vectors be delivered?

Table 14.9   siRNA design guidelines

- **Target selection** should not include the first 50–100 bases downstream of the start codon, the 100 bases upstream of the stop codon or the 5′ and 3′ UTRs (untranslated regions).
- **First nucleotide** – should be either an A or a G.
- **GC content** – between 30% and 55%.
- **siRNA size** – dsRNA < 30 bases to avoid non-specific silencing.

14.4.1 Clinical trials and siRNA

In therapeutic terms, it is still early days for siRNA and hence it is difficult to find information on the products undergoing trials. However, companies such as Sirna do publish their current portfolio with details of the trials and stages reached. Sirna-027, which targets vascular endothelial growth factor receptor 1 (VEGFR-1) and is used for ocular disease, has reached the end of phase1 trials. The Sirna-027 phase 1 clinical trial began in November 2004 with twenty-six patients enrolled in the single-dose escalation trial with six dose cohorts. The trial demonstrated that Sirna's short interfering RNA was safe and well tolerated. Twenty-five of the patients (96%) showed visual acuity stabilisation with 23% of those patients having a clinically significant improvement in visual acuity eight weeks after a single injection. Sirna and its strategic partner, GSK, had two studies with Sirna-027 in the autumn of 2006.[249] The clinical trials identifier is NCT00395057. The same company has a liaison with Allergen to develop siRNA for respiratory diseases. Sirna and Targeted Genetics have a joint development agreement for work in Huntington's disease. Other projects in trials supported by this company target hepatitis B and C, diabetes and, rather strangely, prevention of hair growth.

14.4.2 Is RNAi better than antisense?

Antisense technology has been used to suppress gene expression. Antisense drugs were designed to inhibit gene expression, blocking production of a disease-causing protein. However, initial results with antisense therapies have been disappointing. Natural DNA and RNA have poor pharmacokinetics because they were rapidly degraded in blood and within cells. Thus modifications were made to their structure to improve their stability.

However, the promised specificity of antisense molecules has often been poor and there is evidence that many antisense therapies caused therapeutic responses by non-specific stimulation of the immune system, as opposed to exerting a specific effect on the targeted gene.

Researchers claim that RNA interference has advantages over antisense technology. *In vitro* RNAi inhibits gene expression post-transcriptionally via cytoplasmic mRNA degradation. We can now produce effective siRNAs that appear to give more efficient and longer-lasting effects than antisense oligonucleotides. Indeed, some pharmaceutical companies are claiming silencing with siRNAs at concentrations tens to hundreds of times lower than those used in antisense experiments. However, we must remember that most of the work is experimental, *in vitro*, in cell culture or in animal models and there will be problems associated with the long-term delivery of these molecules.

Points to consider

Look at the list of trials recruiting at www. clinicaltrials.gov
Compare and contrast gene augmentation therapy with gene silencing therapy. Which is the more promising technique?

Notes

223 www.pubmedcentral.nih.gov/articlerender.fcgi?tool=pubmed&pubmedid=9841677

224 http://stemcells.alphamedpress.org/cgi/content/full/18/5/307

225 Verspieren, P. *et al.* An acridine-linked oligodeoxynucleotide targeted to the common 5′ end of trypanosome mRNAs kills cultured parasites. *Gene*, 1987, 61, 307–15

226 www.pubmedcentral.nih.gov/pagerender.fcgi?artinstid=40268&pageindex=1

227 Kipshidze, N. *et al.* Local delivery of c-myc neutrally charged antisense oligonucleotides with transport catheter inhibits myointimal hyperplasia and positively affects vascular remodeling in the rabbit balloon injury model. *Catheter Cardiovasc. Interv.*, 2001, 54(2), 247–56

228 Kipshidze, N. *et al.* Antisense therapy for restenosis following percutaneous coronary intervention. *Expert Opin. Biol. Ther.*, 2005, 5(1), 79–89

229 www.pubmedcentral.gov/picrender.fcgi?tool=pmcentrez&blobtype=pdf&artid=121375

230 www.pubmedcentral.gov/picrender.fcgi?tool=pmcentrez&blobtype=pdf&
     artid=553987
231 Casey, B.P. and Glazer, P.M. Gene targeting via triple-helix formation. *Prog.
     Nucleic Acid Res. Mol. Biol.*, 2001, **67**, 163–92
232 www.rcsb.org./pdb/explore/explore.do?structureId=1BWG
233 Seidman, M.M. and Glazer, P.M. The potential for gene repair via triple
     helix formation. *J. Clin. Invest.*, 2003, **12**, 487–94
234 Grant, K.B. and Dervan, P.B. Sequence-specific alkylation and cleavage of
     DNA mediated by purine motif triple helix formation. *Biochemistry*, 1996,
     35, 12313–9
235 Trojan, L.A. *et al.* IGF-I: from diagnostic to triple-helix gene therapy of solid
     tumors. *Acta Biochim. Pol.*, 2002, **49**, 979–90
236 www.delphion.com/details?pn=WO09106626A2
237 www.bloodjournal.org/cgi/content/full/91/2/371
238 Sioud, M. Ribozymes and siRnas: from structure to preclinical applications.
     Handb. Exp. Pharmacol., 2006, **173**, 223–42
239 Tanabe, T. *et al.* Oncogene inactivation in a mouse model: tissue invasion
     by leukaemic cells is stalled by loading them with a designer ribozyme.
     Nature, 2000, **406**, 473–4
240 Soda, Y. *et al.* A novel maxizyme vector targeting a bcr-abl fusion gene
     induced specific cell death in Philadelphia chromosome–positive acute lym-
     phoblastic leukaemia. *Blood*, 2004, **104**, 356–63
241 www.bioscience.org/1999/v4/d/macphers/macphers.pdf
242 Ngok, F.K. *et al.* Clinical gene therapy research utilizing ribozymes:
     application to the treatment of HIV/AIDS. *Methods Mol. Biol.*, 2004, **252**,
     581–98
243 Dallas, A. and Vlassov, A.V. RNAi: a novel antisense technology and its
     therapeutic potential. *Med. Sci. Monit.*, 2006, **12**, RA67–74
244 Napoli, C. *et al.* Introduction of a chalcone synthase gene into Petunia results
     in reversible co-suppression of homologous genes in trans. *Plant Cell*, 1990,
     2, 279–89
245 MacMorris, M. *et al.* UAP56 levels affect viability and mRNA export in
     Caenorhabditis elegans. *RNA*, 2003, **9**, 847–57
246 Takaku, H. Gene silencing of HIV-1 by RNA interference. *Antivir. Chem.
     Chemother.*, 2004, **15**(2), 57–65
247 www.nature.com/focus/rnai/animations/animation/animation.htm
248 Elbashir, S.M. *et al.* Duplexes of 21-nucleotide RNAs mediate RNA interfer-
     ence in cultured mammalian cells. *Nature*, 2001, **411**, 494–8
249 www.gsk.com/ControllerServlet?appId=4&pageId=402&newsid=786

15

Gene therapy for cancer

Everyone must know someone who has suffered or died from cancer. This places it in a very different category from inherited diseases, which are relatively uncommon. In the developed world few people die of infectious disease and molecular therapies are targeted more towards prevention, in the form of vaccination, than cure. Cancer is simply the uncontrolled growth of poorly differentiated cells. The cancer is usually derived from mutation of a single cell whose progeny form the tumour mass. However, we should contemplate why growth of a clone of cells would lead to the catastrophic disease we recognise as end-stage cancer. Many cancers of solid organs also have the ability to metastasise and travel to other sites, producing secondary tumours. It is this migration that causes problems in conventional therapy. Interestingly the secondaries of a particular cancer are predictable (e.g. breast to spine), suggesting some type of 'homing' or cellular programming.

15.1 What causes cancer?

If I could answer this question I would undoubtedly win the Nobel Prize! The true causes of cancer are not clear, although we know much about the genes that are turned on and turned off during the process. We know, from the study of inherited cancers, that the process involves oncogenes, as and tumour suppressor genes. Oncogenes are mutated or up-regulated,

Molecular Therapeutics: 21st-century Medicine by Pamela Greenwell and Michelle McCulley.
© 2007 John Wiley & Sons, Ltd

increasing cell proliferation, and tumour suppressor genes are down-regulated or mutated, leading to uncontrolled growth. It is also clear that the cell cycle plays an important role in cancers, since it is here that genes are analysed and repaired if necessary. There are many potential cancer-causing genes associated with cell cycle control, such as *Rb* and *p53*. Cancer can also be seen as a disease involving lack of control over apoptosis or programmed cell death. Cancerous cells do not respond to signals that cause them to enter into apoptosis and hence they remain in the system although they should have been removed.

In some cases, cancers could be associated with problems with differentiation signals where cells destined for a specific function are forced down a different differentiation pathway. DNA damage also plays a role; this involves environmental damage and its repair by cellular enzyme systems. An example of this is skin cancer where the environmental agent is UV light. Viruses and bacteria also play a role, for example HPV, HIV, HTLV, *Helicobacter pylori* and hepatitis B, although in some cases we are not sure how the mechanism works. Finally, immune surveillance should, of course, detect problems and enable the body to destroy them. Immune system dysfunction therefore clearly plays a role.

15.2 Cancer: a multifactorial disease

Cancer is often described as a multifactorial process involving genes and environmental triggers. It involves new mutations in the presence or absence of heritable susceptibility. Many textbooks talk about the 'two-hit' theory which suggests that those with an inherited susceptibility are more likely to contract cancer when they encounter the appropriate stimulus. For example, patients with an inherited mutation in the *Rb* gene have few problems; however, if an environmental stimulus mutates their normal gene copy they will develop retinoblastoma. In homozygous normal individuals the chance that an environmental trigger will knock out both *Rb* genes is remote.

The mutation may silence by either mutation or methylation. Mutation 2 will bring about loss of heterozygosity (LOH)

Figure 15.1   The two-hit theory of cancer

Inherited cancers have been investigated as simple models of the development of malignancy. It was originally felt that such studies would shed light on the oncogenic processes not only in inherited but also in spontaneous forms of cancer. However, it quickly became clear that even inherited cancers could arise from a series of mutations. Even a well-studied disease such as the hereditary colon cancer, familial adenomatous polyposis coli (FAP), involves more than four different genetic events and no one is sure of the original trigger. These mutations involve oncogenes and tumour suppressor genes.

Spontaneous tumours are more complex; for example, small-cell carcinoma of the lung has been shown to involve between 10 and 15 mutations. These mutations involve oncogenes, tumour suppressor genes and genes controlling the cell cycle. Clearly it will be difficult to determine which mutation should be treated by gene therapy. If we turn off the oncogenes would the patient recover or would they still have enough mutations to maintain the cancer?

15.3 Cancer statistics

Even seemingly straightforward statistics can be confusing. You may read that colorectal cancer is the second commonest cancer in men in the UK. In another article, you may read that colon cancer is the third commonest cancer in men in the UK, and prostate cancer is the second most common. Both these statements are true. This is because the incidence of colorectal cancer is worked out by adding the incidences of colon cancer and rectal cancer together. If you are trying to raise money for colon cancer research, it is helpful to add the figures together to make colorectal cancer the second commonest. But if you are raising money for prostate cancer research, then you can separate the colon and rectal cancer figures to make prostate cancer come out second! Non-melanoma skin cancer has been excluded from both the examples quoted. It is actually the second commonest cancer overall, after lung cancer.

Non-melanoma skin cancer is often excluded from cancer incidence statistics because, although it is common, it is very easily treated and usually cured. Testicular cancer is sometimes quoted as being the commonest cancer affecting young men in the UK. This is true if you only include men under 50 years of age in your figures. If you are only looking at cancer in children (under-15s), then leukaemia becomes the commonest type of cancer. These figures provide an illustration of the care with which one should interpret results.

15.4 Best treatment currently available

15.4.1 Avoidance

It would seem obvious that, if cancer involves genetic susceptibility and an environmental trigger, if we avoid environmental exposure we will prevent the disease. Clearly this is simplistic! There are instances where avoidance is the best therapy, for example if people refrained from exposing themselves to the sun the incidence of melanoma would drop dramatically. However, every year people ignore warnings of this type. Most lung cancers are the result of smoking or exposure to cigarette smoke, yet people still smoke, despite numerous campaigns. Cancer of the cervix is caused by HPV; the more sexual partners a female has, the more likely she is to meet one with HPV. The answer is simple: use condoms. Yet such measures are rarely used. There is also an increase in colon cancer associated with consumption of red meat and animal fats, though the correlation is weaker than that between smoking and lung cancer, yet many of us will still consume these foods despite the risks.

Despite knowledge, publicity and education it is difficult to persuade people to heed warnings and, although avoidance must play a role, we certainly are not persuading the public to take fewer risks.

15.4.2 Screening

If we can't persuade people to avoid cancer triggers, perhaps we should invest in screening programmes to detect cancers at an early stage and treat them before they metastasise. Many women in the UK are screened for cancer of the cervix. The programme has not been completely successful because of mistakes in interpretation. The value of mammography has also been questioned in breast cancer screening. Current work on prostate cancer states that more men die with prostate cancer than die from it. This suggests that screening programmes may detect disease that will never seriously threaten the health of some of those testing positive. In the USA they have tried methods for screening for blood in faeces as a marker of cancer of the colon. Essentially, a specially treated toilet paper was used that turned blue in the presence of blood. However, more sufferers of haemorrhoids than people with colon cancer were detected. Screening programmes to detect changes in oncogenes and tumour suppressor genes have been suggested, but as yet we do not have the technology to apply these tests to large numbers of people and we are not sure how to interpret the results obtained.

15.4.3 Surgery

Provided that the tumour has not metastasised and is readily accessible, surgery is an excellent approach. However, if metastasis has occurred, removal of the tumour may remove growth inhibitors that have kept secondary tumours in check. There are concerns that surgery itself may liberate cancer cells into the blood stream, causing problems later. If we understood the mechanisms involved in metastasis we could design therapies to limit damage and cure disease. Unfortunately, many cancers are not detected early and have already spread before the patient sees a GP. This is particularly true of some solid tumours, including ovarian and colon cancer.

15.4.4 Chemotherapy

Chemotherapy is usually used in conjunction with surgery to ensure complete removal of the tumour. Most of the active agents used in chemotherapy damage the DNA or the replication mechanism of the cell. Conventional cancer chemotherapy has targeted rapidly dividing cells and therefore should not harm most 'normal' tissues. Sadly, some chemotherapeutic agents affect bone marrow, which proliferates rapidly, and in some cases of breast cancer therapy, bone marrow support must be given to the patients post-chemotherapy. Anti-cancer agents may affect cells in a number of ways: some affect transcription and replication, for example, anthracycline; doxorubicin inhibits topoisomerase II; camptothecin inhibits topoisomerase I and *Vinca* alkaloids and taxol inhibit polymerisation or depolymerisation of tubulin which is essential for spindle formation. Other more old-fashioned drugs are anti-metabolites which prevent cells from, synthesising DNA or replicating; for example, methotrexate blocks dihydrofolate reductase activity.

Chemotherapy is really only suitable for leukaemias, lymphomas and germ cell tumours of the young. Chemotherapy of solid tumours is a problem but promising results have been reported in some trials in ovarian, breast and small-cell lung carcinomas. However, re-emergence of drug-resistant clones is a problem. These cannot be treated with the same therapy and will require a novel approach.

15.4.5 Radiotherapy

Radiotherapy is based on the fact that the treatment causes cell damage that may then resist repair, thus killing the cells. It depends on factors such as pH and oxygenation of the target cells, and many tumours are heterogeneous with respect to these factors. Different cells show different responses to radiation and also vary with respect to their ability to repair damage accurately. In general, normal tissue repairs more quickly than cancerous tissue does and therefore, if repetitive treatments are given, cumulative damage occurs to the tumour cells but not to the 'normal' tissue. This results in specific killing of cancer cells.

15.5 Do chemo- and radiotherapy cause problems?

Chemo- and radiotherapies aim to damage DNA in the cancerous cells. Although we try to target the treatment away from 'normal' tissue, DNA damage will occur, and in the long term can cause tumours related to the treatment, not to the original cancer. However, all therapies come with an attached risk. Even new drugs such as tamoxifen are a problem and carry some risks. During the early use of this drug, five women undergoing treatment died from liver disease, eleven showed liver complications and fifteen showed abnormalities of their blood cells. Nevertheless, we must look at the benefit:risk ratio to determine whether, despite all the associated problems, a treatment will help a patient.

15.6 New cancer therapies

It is clear that we need new cancer therapies. However, there are problems in evaluating potential therapies that may be limiting our detection of new agents. Traditionally, new compounds and therapies are tested first in tissue culture, but researchers have now determined that, whilst this may be useful for detecting agents useful for treatment of leukaemias, it fails to detect some treatments that might be useful in other tumours. It is near impossible to determine in a tissue culture model whether a treatment can be delivered into a solid tumour, and metastasis cannot be modelled in culture. Additionally, culture models use clones of cells, whereas the human body is made up of different cell types, and treatments active on one cell type may be ineffective on another.

15.7 Cancer models in animals

There is some concern that animal models for cancer are poor in that the cancer-causing genes frequently have a different effect in models than they do in humans. This then makes it difficult to determine the likelihood of the success of such gene-directed therapies. Additionally, we know that in humans most cancers are a multi-stage event involving five or more genes. Which should we target?

Cancer therapies have also been tested in cultured cells. For example, researchers have turned off the *bcr-abl* transcript in leukaemic cells and demonstrated that this gives a significant growth advantage to normal cells. This is really seen to be useful not in treating the patient directly but in, for example, bone marrow purging. In this case autologous marrow could be harvested purged with antibodies and then incubated with anti-sense oligonucleotides to ensure that most of the cells growing were *bcr-abl*-negative.

All human therapies must be validated in animal models. However, in animal models for cancer the cancers are induced artificially either by drugs or the introduction of cancer cells into the animal. These mechanisms do not reflect the mechanisms of carcinogenesis seen in humans, often the tumours are not allowed to metastasise and animal models are poor models with respect to their immunological response. Animals have different chromosome numbers and gene arrangements and may respond differently to anti-cancer drugs than do humans. They also have a shorter lifespan; so it is difficult to assess the results when for example a rat is cured of a cancer in two months: will a human being treated in the same way have two months' cure or is it related to the overall projected lifespan, e.g. 2 months × 70/3 (average age of death of a human/average age of death of rat). Many treatments work in mice but not in humans.

15.8 What kinds of gene therapy can we use to treat cancer?

Delivery of genes such as tumour suppressor genes will attract the same problems as we have previously discussed in gene augmentation therapy. However, in this case we may be able to harness retroviruses to target the rapidly dividing tumour cells, or use adenoviruses as a single-dose therapy, hence bypassing problems of immunogenicity. Nevertheless, therapy must be targeted, as we would need to avoid the side effects of delivering extra copies of genes involved in the cell cycle to normal cells. All gene therapies for cancer are currently experimental. Some are done only in animals, others are in trials in a limited number of patients.

Researchers recognise that the major problems associated with this type of therapy must be addressed before gene therapy can be used successfully. The problems are listed in Table 15.2. Probably the single most important point is that to effect a long-term cure all the cancer cells must be destroyed, and many workers believe that this can only be achieved by using a combination of traditional and newer methods. From the gene therapist's point of view we need to identify our target gene, which ideally should be the primary cause of the tumour. However, are we sure that removing the primary cause will indeed cure the patient? We need to target cancer cells specifically and therefore some type of receptor targeting or cell-specific promoter will be required. Lastly, we need to deliver the DNA – currently a major stumbling block, but possibly easier if a one-off treatment will be sufficient.

Table 15.1   Molecular strategies for cancer therapy

- deliver tumour suppressor gene, turn on tumour suppresser gene or repair mutation
- turn off oncogene or repair mutation
- deliver suicide genes
- add drug-resistance genes to normal tissues

Table 15.2   Perceived problems in cancer gene augmentation therapy

- ideally target primary cause
- need to identify target
- need to deliver genes successfully
- need to target expression only to the cancer cells
- need to completely remove the cancer for long-term cure

15.9 Experimental strategies for gene augmentation therapy

The most obvious strategies for cancer therapy would be the delivery of tumour suppressor genes that in theory would inhibit the growth of cancer cells. A number of clinical trials, mainly phase 1 and 2, have looked at the delivery of the tumour suppressor gene *p53* for the treatment of, for example, oesophageal squamous cell carcinoma,[250] small-cell lung carcinoma,[251] ovarian cancer[252] and head and neck cancer.[253] As yet, these therapies have not been tested on a large enough cohort for us to be sure of the effect, although in each trial, some patients gained benefit.

15.9.1 Killing cells with ganciclovir or suicide therapy

This is an elegant approach that relies on the delivery of the *thymidine kinase (TK)* gene to rapidly dividing cancer cells. Such delivery may be accomplished using a retrovirus. Cells that take up the viral vector produce thymidine kinase and, when the patient is treated with the drug ganciclovir (most often prescribed for cytomegalovirus infection), the enzyme metabolises the drug to yield toxins that kill the cell. Cells that do not take up the vector remain unharmed. This approach has been used in the treatment of brain, ovarian, head and neck tumours and melanomas. In the case of brain tumours, the viral vectors are delivered into the brain in a packaging cell line that produces retroviruses *in situ*. In other forms of tumour, researchers inject retroviral vectors directly into tumour mass.

The results were spectacular in animal models and the therapy was then tested in 15 patients with inoperable brain cancers. Of these, five lesions in four patients showed response, three partial and two complete. One of the patients lived for 24 months when they had been expected to die within 6 months. Nevertheless, we should be cautious about interpretation. One of the workers in this field stated, 'I don't want to mislead people by saying this is going to be a cure for brain cancer but I do believe it will be another tool in treatment.' On CAT scans the tumours of treated patients clearly did shrink, but they were not completely removed. Nevertheless, surgeons stated that if a large inaccessible tumour was shrunk in this way it could then make it amenable to surgery that would otherwise not have been an option. However, in 2000, results from a large phase III trial showed that the therapy did not give significant effects.[254] Since that time, there have been suggestions that commonly used drugs such as dexamethasone may adversely influence the therapy.[255] However, other trials have shown partial responses.[256]

In animal trials on head and neck cancer three out of four animals treated with this technique showed complete remission and two of four were still tumour-free after 60 days. Again, this is not a cure, but ovarian and head and neck cancers are very aggressive and spread quickly and are difficult to treat by conventional methods. Any therapy that slowed the progression of such a cancer would give surgeons time to look at other options. However, there is no evidence that trials are under way in humans.

A strategy involving the same HSV–TK system has been used in HPV infection.[257] In this case, the HPV E2 protein regulated the TK expression. Thus, in cells which were HPV-negative, no gene product was formed, whereas in HPV-positive cells TK was expressed and rendered the cells susceptible to ganciclovir.

15.9.2 Prodrug activation therapy[258]

This type of gene therapy is similar to the HSV–TK approach. Here, gene therapy is used alongside drug therapy to enable scientists to specifically deliver large doses of drugs directly to tumours. Two approaches have been described, VDEPT (virally directed enzyme prodrug therapy) and GPAT (gene prodrug activation therapy). Essentially both rely on the insertion into the cancer cell of a gene that will convert a harmless drug precursor into a fully active drug. An example of such a therapy is MetXia-P450, a novel recombinant retroviral vector that encodes the human cytochrome P450 type 2B6 gene (CYP2B6). Cytochrome P450 enzymes are normally expressed in the liver and convert the prodrug cyclophosphamide to an active phosphoramide mustard and acrolein. Thus, gene therapy to deliver CYP2B6 to the tumour results in local prodrug activation and increased concentrations of the active drug in the cancerous cells.[259] The trial involved 12 cancer patients, 9 with breast cancer and 3 with melanoma. Preliminary results suggested an anti-tumour response in two patients: partial response in one patient and stable disease in one patient. In a similar approach,[260] adenovirus-encoding nitroreductase was administered to patients with operable liver cancer to activate CB1954 (5-(aziridin-1-yl)-2,4-dinitrobenzamide) to a short-lived, highly toxic DNA cross-linking agent. However, in this case, although the vector was well tolerated with minimal side effects, it had a short half-life in the circulation, and stimulated a strong antibody response.

One group[261] has devised a 'double suicide' gene therapy involving administration of HSV–TK and cytosine deaminase in an attempt to treat locally recurrent prostate cancer. The therapy was designed to make the patients responsive to drug therapies with 5-fluorocytosine and ganciclovir and to radiation therapy. Following treatment of 16 patients, 2 were negative for adenocarcinoma at 1-year follow-up.

Clearly there will be problems of delivering the genes to cancer cells and a worry that, if not all cancer cells are infected, then the treatment might be unsuccessful. Currently most of this work is in cultured cells or in animal models.

A similar approach has been used to kill colon tumour cells. The product of the *DeoD* gene converts non-toxic deoxyadenosine analogues to toxic purines. In cell culture, cells were infected with virus-carrying *DeoD* and then exposed to deoxyadenosine analogues. The infected cells died and so did their neighbours by virtue of the bystander effect. However, problems of delivery have prevented use of this therapy in humans.[262]

As yet, therefore, the usefulness of this technology is still unclear, delivery and expression are still problems and promising early trials have not led to breakthroughs in treatment.

15.9.3 Enhancing the immune system with gene therapy

Work in the USA led to another well-publicised failure in the field of gene therapy. Here is another example of a treatment that worked so well in animal models that researchers were given permission for one of the largest gene therapy trials. Sadly, the treatment had no effect in humans, but the results were not published – such a publication would have severely affected future funding.

In essence the work centred on the delivery of TNF (tumour necrosis factor) by tumour-infiltrating lymphocytes (TILs) engineered to carry the *TNF* gene. TILs would target cancer cells, TNF would kill them. In animals that is exactly what happened, but in humans there was no response and the treatment proved toxic.

Nevertheless, other groups have devised novel therapies based on the fact that although cancer cells are non-self, in that they express inappropriate antigens, they are frequently ignored by the host immunosurveillance. This may be because the antigens are present at low levels or are in some way hidden. In animal models, treatment of cancer cells, which do not elicit immune response with IL-2, has been shown to result in increased presentation of the antigens to the immune system, leading to an immune response. Once the immune system recognises the antigens presented in this way it will also recognise the cancer cells themselves. However, we should recall that therapies involving interleukins have been shown to be toxic.

In a study of therapy for renal carcinoma, scientists experimented on cell lines in culture and delivered DNA-encoding IL-2 production, linked to a lectin (sugar-binding protein). The lectin recognised cell surface receptors and the *IL-2* gene was taken up and IL-2 expressed. If the original cell lines were injected into an animal model they elicited no immune response whereas after transfection with the *IL-2* gene the cells were recognised by the immune response of the animal. This led scientists into clinical trials. The idea of the experiment was to take renal tumour cells from the patient, transfect *ex vivo* and then return the cells to the patient with the aim of provoking an immune response. The method was used in patients with terminal kidney cancer. Although 15% of the 225 patients in the trials had tumour shrinkage, the treatment was shown to be toxic to the heart and other organs. None of the patients treated was cured and the development costs of this therapy have been estimated at $100 million.

In a separate trial on prostate cancer,[263] IL-2 was delivered using liposomes. The authors reported that transient decreases in serum prostate-specific antigen (PSA) were seen in 16 of 24 patients (67%) on day 1. The response was still detectable at day 8 in 14 of the patients. However, in 8 patients the PSA level rose. This phase I trial demonstrated the safety of intra-prostatic delivery of IL-2 with transient responses seen after therapy.

Intra-lesional delivery of cytokine genes has been used as a therapeutic strategy for the treatment of head and neck cancer.[264] The researchers hypothesised that a therapy could be designed that would provide anti-tumour activity by expression of interleukin (IL)-2 from a recombinant plasmid which would lead to induction of endogenous interferon (IFN)-gamma and IL-12 cytokines by immunostimulatory DNA. Gene delivery was by virtue of cationic liposomes to the floor of the mouth of tumour-bearing mice. The study was extended to a 10-patient phase I study. This was followed by two large-scale phase II multi-institutional and multi-national studies. In mouse models, the delivery of cationic liposomes containing the *IL-2* plasmid significantly inhibited tumour growth and production of endogenous IFN-gamma and IL-12, but not IL-2, were induced. The phase I human trial demonstrated safety. Phase II studies determined efficacy and feasibility of this therapy as a single or combination therapy for head and neck cancer.

Researchers have investigated the potential utility of delivery of a combination of *MUC-1* and *IL-2*[265] genes. MUC-1 is a highly glycosylated mucin that is over-expressed and aberrantly glycosylated in many tumour cells. This results in exposure of cryptic epitopes which may play a role in tumour immunity. A phase I clinical trial was designed to determine the maximum tolerated dose, safety of a multiple-dose regimen, and the immunologic effect of vaccinia virus expressing *MUC-1* and *IL-2* genes in patients with advanced prostate cancer. No high-grade toxicity was observed but clinical response was observed after the fourth injection in only one patient. In this patient there was evidence that immune stimulation had occurred. This single result suggests that the therapy had effect, but further work needs to be carried out to determine whether the results are clinically significant and warrant further trials.

Following a pilot clinical study of combined *IL-2/HSV-TK* gene therapy for recurrent glioblastoma multiforme (GBM),[266] a total of twelve patients were treated by intra-tumour injection of retroviral vector-producing cells followed by intravenous ganciclovir. A marked and persistent increase of intra-tumour and plasma Th1 cytokine levels was demonstrated after injection. During magnetic resonance imaging, two patients showed a partial response, including a patient in whom there was disappearance of a distant tumour mass, four had a minor response, four had stable disease, and two had progressive disease. The 6- and 12-month progression-free survival rates were 47% and 14%, respectively. The 6- and 12-month overall survival rates were 58% and 25%, respectively. This method appeared to be safe and provided effective transfer of the therapeutic genes into the target tumour cells which activated a systemic cytokine cascade, with tumour responses seen in 50% of cases. It is, however, difficult to judge the outcome compared to other treatments. Conventional approaches use surgery followed by both radiation treatment and chemotherapy. In some cases, chemotherapeutic wafers are used to kill any remaining tumour cells. However, even following the triad of surgery, radiation and chemotherapy, the average survival is approximately 12 months.

15.10 Gene silencing technologies and cancer

All forms of gene silencing techniques have been trialled for cancer therapy. Some have targeted oncogenes, others growth factor genes and angiogenesis. Real success has not yet been reported. Table 15.3 shows a range of targeted genes for siRNA and it is clear that every mechanism seen in cancer has been the target of silencing technologies.

Table 15.3   siRNAs in cancer gene therapy[267]

Gene	Cancer	Effect
Bcr-Abl	Chronic myeloid leukaemia (CML) and distinct variants of acute lymphoblastic leukaemia (ALL)	Inhibition of cell growth, induced differentiation, increased sensitivity to γ-irradiation and imatinib mesylate
K-ras	Pancreatic cancer, leukaemia, thyroid carcinoma, lung and colorectal	Reduced proliferation and angiogenesis, enhanced radiation sensitivity, loss of tumorigenicity
PKA RIα	Pancreatic, breast	Induction of apoptosis and decrease of cell proliferation
HER-2/neu (c-erbB2)	Breast, ovarian, colon and gastric	Cell growth arrest, increased expression of anti-angiogenic factors, induction of apoptosis
EGFR (erbB1)	Breast, prostate, ovarian and gastric	Suppression of tumour angiogenesis and growth, induction of apoptosis
VEGF	Breast, prostate, ovarian and gastric	Suppression of tumour angiogenesis and growth
Bcl-2	Breast, lung, prostate, pancreatic cancer	Anti-proliferative activity through the induction of apoptosis
c-*myc*	Breast, prostate, gastrointestinal cancer, lymphoma, melanoma, myeloid leukaemia	Induction of apoptosis, decrease in cell proliferation rate
MDR1	Breast cancer	Enhance effects of chemotherapeutic agents
PTEN/PI3K	Colon, breast, prostate, ovarian and gastric	Enhance effects of chemotherapeutic agents, suppression of tumour growth, induction of apoptosis
Survivin	Pancreatic, colorectal	Enhanced radiation sensitivity

It is clearly premature to assess the success of siRNA since few drugs have actually reached the marketplace. However, in 2003, Genta, the producers of Genasense (G3139 or oblimersen sodium), an antisense molecule targeting the *bcl-2* gene, submitted a drug application to the FDA. Genasense was said to inhibit chemotherapy-induced apoptosis and enhance response of patients with melanoma to alternative therapy. In the trials Genasense, in combination with dacarbazine (DTIC) was compared to treatment with dacarbazine alone in patients with metastatic melanoma.

Of 771 patients enrolled, those who received Genasense in addition to dacarbazine had a 51% improvement in median progression-free survival (74 days vs. 49 days), an improvement in durable response rate at six months (13 patients vs. 5 patients), and a 72% increase in overall anti-tumour response rate (11.7% vs. 6.8%). At follow-up, 6 additional patients achieved complete responses with Genasense, but none did in the dacarbazine-alone group.

On the face of it, this may seem like a worthwhile treatment. However, common serious adverse events with Genasense have been fever (5.9% vs. 3.1%), neutropenia (21.3% vs. 12.5%), thrombocytopenia (15.6% vs. 6.4%), leukopenia (7.5% vs. 3.9%), anaemia (7.0% vs. 4.7%) and nausea (7.0% vs. 2.5%).[268] In early 2004, it was determined that the response rate and progression-free survival did not outweigh the increased toxicity. Despite the problems shown with melanoma treatment, this drug is also being trialled in conjunction with Fludarabine and Rituximab in subjects with chronic lymphocytic leukaemia. Additional trials have featured another antisense molecule targeted to bcl-2, SPC2996.[269]

An antisense compound in phase II development by Lorus Therapeutics, GTI-2501, contains sequences directed against R1 and R2 components or ribonucleotide reductase. This is being tested in combination with GTI-2501 and docetaxel in patients with hormone refractory prostate cancer (HRPC).[270]

The future for RNAi is difficult to define as no RNAi-based product is in clinical development yet. As yet there have been no major safety concerns and as the clinical trials are just starting, regulations are in preliminary stages. The value of the drug discovery market based on RNAi is suggested to be $650 million in 2005, increasing to $1 billion in 2010 and $1.5 billion in 2015. Even if only a few products reach the market by the year 2010, revenue from sales of RNAi-based drugs will increase to $3.5 billion and to $5.9 billion in 2015. Approximately 18 of over 100 companies involved in developing RNAi technologies are developing RNAi-based therapeutics.[271,272]

15.11 Conclusion

There are many ideas about the types of cancer therapies that could be used in therapy: however, as yet these have not yet come of age. More research and well-controlled clinical trials are needed to prove the efficacy of the products. However, molecular therapeutics is unlikely to lead to a 'one-off' cure for cancer: we will simply develop another tool in our arsenal to fight the disease.

Points to consider

Compare and contrast the types of therapy used in cancer treatment. Which is the best?

Why is cancer easier than inherited diseases to treat with gene therapy?

Notes

250 Shimada, H. *et al.* Phase I/II adenoviral p53 gene therapy for chemoradiation resistant advanced esophageal squamous cell carcinoma. *Cancer Sci.*, 2006, 97, 554–61

251 Fujiwara, T. *et al.* Multicenter phase I study of repeated intratumoral delivery of adenoviral p53 in patients with advanced non-small-cell lung cancer. *J Clin. Oncol.*, 2006, 24, 1689–99

252 Wolf, J.K. *et al.* A phase I study of Adp53 (INGN 201; ADVEXIN) for patients with platinum- and paclitaxel-resistant epithelial ovarian cancer. *Gynecol. Oncol.*, 2004, 94, 442–8

253 Zhang, S.W. *et al.* Treatment of head and neck squamous cell carcinoma by recombinant adenovirus-p53 combined with radiotherapy: a phase II clinical trial of 42 cases. *Zhonghua Yi Xue Za Zhi.* 2003, 83, 2023–8

254 Rainov, N.G. A phase III clinical evaluation of herpes simplex virus type 1 thymidine kinase and ganciclovir gene therapy as an adjuvant to surgical resection and radiation in adults with previously untreated glioblastoma multiforme. *Hum. Gene Ther.*, 2000, 11, 2389–401

255 Robe, P.A. *et al.* Dexamethasone inhibits the HSV-tk/ ganciclovir bystander effect in malignant glioma cells. *M. C. Cancer*, 2005, 5, 32

256 Colombo, F. *et al.* Combined HSV-TK/IL-2 gene therapy in patients with recurrent glioblastoma multiforme: biological and clinical results. *Cancer Gene Ther.*, 2005, 12, 835–48

257 Sethi, N. and Palefsky, J. Treatment of human papillomavirus (HPV) type 16-infected cells using herpes simplex virus type 1 thymidine kinase-mediated gene therapy transcriptionally regulated by the HPV E2 protein. *Hum. Gene Ther.*, 2003, 14, 45–57

258 Xu, G. and McLeod, H.L. Strategies for enzyme/prodrug cancer therapy. *Clin. Cancer Res.*, 2001, 7, 3314–24

259 Braybrooke, J.P. *et al.* Phase I study of MetXia-P450 gene therapy and oral cyclophosphamide for patients with advanced breast cancer or melanoma. *Clin. Cancer Res.*, 2005, **11**, 1512–20

260 Palmer, D.H. *et al.* Virus-directed enzyme prodrug therapy: intratumoral administration of a replication-deficient adenovirus encoding nitroreductase to patients with resectable liver cancer. *J. Clin. Oncol.*, 2004, **22**, 1546–52

261 Freytag, S.O. *et al.* Phase I study of replication-competent adenovirus-mediated double suicide gene therapy for the treatment of locally recurrent prostate cancer. *Cancer Res.*, 2002, **62**, 4968–76

262 Sorscher, E.J. *et al.* Tumor cell bystander killing in colonic carcinoma utilizing the *Escherichia coli DeoD* gene to generate toxic purines. *Gene Ther.*, 1994, **1**(4), 233–8

263 Belldegrun, A. *et al.* Interleukin 2 gene therapy for prostate cancer: phase I clinical trial and basic biology. *Hum. Gene Ther.*, 2001, **12**, 883–92

264 O'Malley, B.W. *et al.* Combination nonviral interleukin-2 gene immuno-therapy for head and neck cancer: from bench top to bedside. *Laryngoscope*, 2005, **115**, 391–404

265 Pantuck, A.J. *et al.* Phase I trial of antigen-specific gene therapy using a recombinant vaccinia virus encoding MUC-1 and IL-2 in MUC-1-positive patients with advanced prostate cancer. *J. Immunother.*, 2004, **27**, 240–53

266 Colombo, F. *et al.* Combined HSV-TK/IL-2 gene therapy in patients with recurrent glioblastoma multiforme: biological and clinical results. *Cancer Gene Ther.*, 2005, **12**, 835–48

267 Rychahou, P.G. *et al.* RNA interference: mechanisms of action and therapeutic consideration. *Surgery*, 2006, **140**, 719–25

268 www.cancer.gov/search/ViewClinicalTrials.aspx?cdrid=360621&version=HealthProfessional&protocolsearchid=2976732

269 www.cancer.gov/search/SearchClinicalTrialsAdvanced.aspx

270 www.centerwatch.com/professional/cwpipeline/eyeon_prostate.html

271 www.bioportfolio.com/LeadDiscovery/PubMed-120402.html

272 http://www.leaddiscovery.co.uk/reports/pharmaceutical-reports-markets-and-pipelines/rnai_technologies_markets_and_companies.html

16

Single-nucleotide polymorphisms (SNPs) and therapy

The host response to disease and disease treatment varies. This variation may make some people, or races, more susceptible to infection or cancer or make them metabolise drugs differently. In this chapter we are mainly concerned with the effect of SNPs (single-nucleotide polymorphisms) on drug therapy and the development of designer drugs. SNPs are apparently benign differences seen between genomes. A definition of a polymorphism would be 'the simultaneous occurrence in the population of allelic variations'; additionally the allelic variant should be present in more than 1% of the population. We need to be careful about the definition as, for example, in some populations of African descent 1 in 20 (5%) of individuals carry the gene for sickle cell disease. Thus, in general polymorphisms are defined as mutations that do not directly cause disease.

As an example, we can look at the blood group *ABO* polymorphism. There are three alleles, *A*, *B* and *O*; although there is no direct threat to health by virtue of inheriting any of the polymorphic genes, blood group A individuals are more likely to suffer from cancer and heart disease whereas blood group O are more likely to have ulcers and blood group B to catch typhoid. We cannot in this situation use polymorphisms as diagnostic tools, since we are not suggesting that all, or only, blood group A individuals will die of cancer, we are simply implying that blood group As have twice the risk of contracting cancer than blood group O individuals.[273]

There are situations where SNPs are good prognostic markers. SNPs do not directly cause disease, but they can help determine the likelihood that

Molecular Therapeutics: 21st-century Medicine by Pamela Greenwell and Michelle McCulley.
© 2007 John Wiley & Sons, Ltd

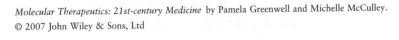

someone will develop a particular disease. The *ApoE* gene, which encodes apolipoprotein E, is a good example of how SNPs affect disease development. *ApoE* is one of the genes associated with Alzheimer's disease. It contains two SNPs that result in the production of three alleles: *E2, E3* and *E4*. An individual who inherits at least one *E4* allele will have a greater chance of getting Alzheimer's disease. Individuals inheriting the *E2* allele are significantly less likely to develop the disease.[274] Once again, these are not absolute markers.

SNPs can also be used in linkage studies of disease. In the past, such studies involved finding polymorphic markers along different chromosomes and then, in family studies, determining which markers were inherited with the disease gene. The nearer the marker to the gene, the less likely they were to be separated from each other by crossing-over events (segregation). The number of polymorphisms known and mapped was comparatively low until the inception of SNP projects, such as The SNP Consortium,[275] which had mapped more than 1.5 million SNPs in the human genome by the end of 2001. This has enabled us to use SNPs as markers for linkage studies rather than other polymorphic genes. This type of strategy has been used to determine genes linked to a number of diseases, including hypertension.[276]

Linkage and predisposition studies are really diagnostic, not therapeutic, issues. However, in the metabolism of drugs, there are genes in which polymorphisms affect the way in which drugs are methylated, absorbed, transported and cleared from the system. This makes SNPs an important area of study when developing therapeutic strategies. In 2002, Klaus Lindpaintner, vice-president of research and director of Roche Genetics and Roche Center for Medical Genomics, said that they viewed *'genetics as one tool in the toolkit, no more than that, so we are investing cautiously. It's currently no more than 5 percent of our total R&D expenditure, but that's no pittance, either.'*

Currently, if you have high blood pressure your GP may prescribe an ACE (angiotensin-converting enzyme) inhibitor drug at a low concentration. If you do not respond, they may increase the dose. However, they will know from experience that some patients will never respond to that drug and until recently the reason was unclear. Pharmaceutical companies have previously been limited to developing agents to which the 'average' patient will respond. As a result, many drugs that might benefit a small number of patients never make it to market. However, in the last two decades, SNPs have been found in a range of genes involved in drug metabolism and it is now becoming clear that SNP analysis will be useful in determining and understanding why individuals differ in their abilities to absorb or clear certain drugs and why some individuals experience adverse side effects to a particular drug.

SNPs have been shown to be associated with the absorbance and clearance of therapeutic agents. SNPs can change the structure and activity of enzymes involved in prodrug activation or drug clearance. Changes in these functions result in differences in the ways these drugs are metabolised in different individuals. Thus, if a drug needs to be metabolised to become active, an SNP which induces rapid metabolism leads to good efficacy and rapid effect, whereas one which induces poor metabolism will lead to poor efficacy and accumulation of the prodrug with potential side effects. If, on the other hand, the active drug needs to be metabolised to clear it from the system, the poor metaboliser could show good efficacy of the drug but drug accumulation would occur with potential hazards; thus a lower dose would be better in this case. On the other hand in the rapid metaboliser, the drug could be cleared too soon and offer little benefit and hence a greater dose may be required.[277]

Table 16.1 Drug metabolism and SNPs

Drug	Rapid metaboliser	Poor metaboliser
Prodrug needs to be metabolised to give effect, e.g. codeine	Good efficacy and rapid effect	Poor efficacy and prodrug accumulation
Active drug needs to be metabolised to be inactivated	Poor efficacy and greater dose required	Good efficacy but accumulation of active drug can occur, therefore may need lower dose

We know that SNPs play a role in the metabolism and functioning of many drugs and Table 16.2 outlines, for a small number of common drugs, the numbers of patients who either do not respond or who are poor responders. It is clear that this is a major problem, resulting in GPs prescribing drugs that in many people have no benefit. This has three effects:

- the patient does not get better despite treatment
- the patient has side effects to the drug
- the health service wastes money.

Table 16.2   Patient response to a range of common drugs

Type of drug	Refractory or poor responders	Uses
ACE inhibitors	10–30%	Prevention of cardiovascular disorders. Congestive heart failure. Hypertension. Left ventricular dysfunction. Prevention of nephropathy in diabetes mellitus.
Beta blockers	15–25%	Hypertension. Angina. Mitral valve prolapse. Cardiac arrhythmia. Congestive heart failure. Myocardial infarction.
Tricyclic anti-depressants	20–50%	Treatment of clinical depression, neuropathic pain, nocturnal enuresis and ADHD (attention-deficit hyperactivity disorder).
Beta-2-agonists	40–70%	Muscle relaxants

Table 16.3   Side effects seen on administration of anti-depressants

Drug	Example	Adverse side effects
Tricyclics	Imipramine	Urine retention in men, dry mouth, tremors, indigestion, cardiac related problems, low blood pressure
Monoamine oxidase inhibitors	Phenelzine sulphate	Hypomania, low blood pressure, tremors, agitation, insomnia
Serotonin re-uptake inhibitors	Fluoxetine (Prozac)	Tremors, genito-urinary tract problems, sexual dysfunction, headache, insomnia
Atypical	Bupropion	Seizures, agitation, tremors, insomnia

If we look at a single class of drugs, the anti-depressants, we can see that there are many side effects of the drugs in some patients. Some of these are potentially life-threatening, others will simply make the depression worse rather than better.

In the *Psychiatric Times* in July 2005, there was a report that the CYP 2D6 poor-metaboliser phenotype was associated with adverse drug reactions to a drug called risperidone (Risperdal), an antipsychotic medication for schizophrenia treatment, while the CYP 3A5 poor-metaboliser phenotype was not significantly associated with the adverse response.[278]

In a trial of 325 patients who were stabilised on risperidone therapy, 73 had moderate-to-marked adverse drug reactions such as resting tremor, stiffness, increased production of saliva, and sedation; 81 patients stopped taking the drug because of adverse drug reactions. The authors found that the CYP 2D6 poor-metaboliser phenotype increased the odds of having moderate adverse drug reactions by 3:1. However, there were patients with the poor-metaboliser phenotype who had no adverse side effects. This is an example of increased risk associated with phenotype rather than an absolute effect.

In HIV, SNPs play a number of important roles. The chemokine receptor CCR5 is the co-receptor through which HIV enteres its target cells. The best-characterised *CCR5* polymorphism for a 32-base-pair deletion within the promoter region (CCR5 D32) which is associated with decreased susceptibility to HIV infection and disease progression. However, other SNPs within the *CCR5* promoter are linked to increased likelihood of HIV infection and progression. A C-to-T change at amino acid 280 (T280M) of the *CX3CR1* sequence has been liked to a more rapid progression to AIDS. However, the authors reported that the association was good for some, but not all populations.[279,280] Different alleles of *HLA* genes also affect the action of anti-virals, for example, with abacavir, an anti-HIV drug, there is a 6 or 7 per cent incidence of serious adverse hypersensitivity in patients with the HLA-B*5701 allele. Should we test patients for the SNP before commencing treatment?

So how does knowledge of SNPs and their association with adverse drug effects and metabolism help in the treatment of disease? Simplistically, we could test for known variants prior to prescription of drugs, though this could be extremely expensive. The realisation that many of the SNPs are present only in specific racial groups has led to speculation about designing drugs not for individuals, but based on racial background. For example, racial differences in SNP frequency in the *methylenetetrahydrofolate reductase* gene have been shown to affect the response of patients with rheumatoid arthritis to methotrexate.[281]

There has been much speculation as to the public response to designing drugs based on racial background, but provided there is sufficient education of the public there should be no more problem than there has been with using genetic tests for sickle cell disease only on those with African or Mediterranean heritage. However, with the increasing numbers of 'mixed-race' relationships, care must be taken to accurately determine genetic background. For example, a patient may phenotypically seem to be white Caucasian, but could have a parent or grandparent who was West of Indian or African descent. Race is a sensitive issue and questions have been raised as to whether it is 'politically correct' to treat by race. However, in the case of some SNPs linked to adverse drug reactions, it is a common-sense approach.

Table 16.4   Racial distribution of poor and rapid metabolisers

Enzyme	% poor metabolisers	% rapid metabolisers
CYP 2D6	8% Caucasians, <1% Orientals, 1.6% African American	5% Europeans
CYP 2C19	3% Caucasians, 20% Orientals	
CYP 2C9	25% Caucasians	
Acetylation	55% Caucasians, 45% African Americans, 10% Japanese, >90% in some Mediterraneans	68% Japanese

Another way of overcoming the cost of testing would be to carry out 'abbreviated SNP profiles' targeted at those genes known to be involved in drug metabolism. During large phase II clinical trials designed to assess drug efficacy, SNP profiles could be identified and used to select patients likely to benefit for phase III trials. This in turn could allow researchers to carry out phase III trials in a more focused way. It would also be possible to undertake surveillance strategies after drug licensing to determine whether those with uncommon serious adverse drug reactions had specific SNPs, and thus to modify the criteria for drug prescription. The alternative would be to test all those receiving the drug for the appropriate SNP but the cost involved would be a factor. There are fears, however, that for some individuals or racial groups specially designed drugs will be required which may not be developed, given either the small numbers requiring the drug or the lack of money to buy the drug within a community.[282]

In 2004, researchers developed a novel ABL-kinase inhibitor for individuals who had developed resistance to the commonly used drug imatinib (Gleevec). Tumour cells can develop SNPs within the BCR-ABL-kinase, which interfere with the drug-binding mechanism. Modelling the wild type and mutant proteins enabled scientists to understand the steric hindrance that prevented binding to the mutant and to develop a new and very promising drug for treatment of patients with resistance to imatinib.[283]

This is an excellent example of the way in which proteomics can be used to facilitate drug design.

In 2002, scientists reported the development of the HapMap:[284] '*a catalogue of common genetic variants that occur in human beings. It describes what these variants are, where they occur in our DNA, and how they are distributed among people within populations and among populations in different parts of the world.*' The International HapMap Project was set up to provide information that could be used to determine which genetic variants are associated with the risk for specific illnesses, with the hope that this would lead to the development of new methods of diagnosing, preventing and treating disease.

It may be, in the future, that microarrays ('gene chips') will be used to determine SNPS and will allow individuals to have their own SNP profile on record. Evidence suggests that this could be useful in safe prescription of medicine, development of new drugs and estimation of risk for diseases.[285,286] However, a note of caution may be raised since these profiles could act as surrogate fingerprints that would allow us to identify individuals.

Clearly, SNPs are important in many areas of bioscience. As the detection technologies develop, we could all have our own SNP profile available, but perhaps we should consider what information could be generated from such data, for example race or susceptibility to disease. Thus strict regulations will need to be followed to prevent misuse of data and address issues of data protection.

Points to consider

Explore the websites below, which provide information about SNPs and their association with disease, frequency within populations and effect of encoded proteins. The completed version of the HapMAp is also available on-line.

www.ncbi.nlm.nih.gov/projects/SNP/
www.hapmap.org/
http://snpper.chip.org/
http://snp500cancer.nci.nih.gov/home_1.cfm?CFID=1642850&
    CFTOKEN=28624924

Notes

273 Greenwell, P. Blood group antigens: molecules seeking a function? *Glyco-conjugate Journal*, 1997, **14**, 159–73

274 www.ornl.gov/sci/techresources/Human_Genome/faq/snps.shtml

275 http://snp.cshl.org/

276 http://hyper.ahajournals.org/cgi/content/full/39/2/323

277 de Leon, J. *et al.* Clinical guidelines for psychiatrists for the use of pharma-cogenetic testing for CYP450 2D6 and CYP450 2C19. *Psychosomatics*, 2006, **47**, 75–85

278 http://www.practicalpsychiatry.com/pt/re/jpsychpract/abstract.00131746-200507000-00004.htm;jsessionid=G2KhDyQHNltSGJgL4v18VyJrJWwml1 PvQTf5LFhsNDsc4nJyHJzJ!1330140564!181195629!8091!-1

279 www.natap.org/2004/HIV/062804_02.htm

280 www.iasusa.org/pub/topics/2005/issue3/90.pdf

281 Beasley, T.M. *et al.* Racial or ethnic differences in allele frequencies of single-nucleotide polymorphisms in the methylenetetrahydrofolate reductase gene and their influence on response to methotrexate in rheumatoid arthritis. *Ann. Rheum. Dis.*, 2006, **65**, 1213–18

282 Brazell, C. *et al.* Maximizing the value of medicines by including pharmaco-genetic research in drug development and surveillance. *Br. J. Clin. Pharma-col.*, 2002, **53**, 224–31

283 Shah, N.P. Overriding imatinib resistance with a novel ABL kinase inhibitor. *Science*, 2004, **305**(5682), 399–401

284 www.hapmap.org/whatishapmap.html.en

285 www.biomedcentral.com/1471-2334/6/82

286 www.pubmedcentral.nih.gov/articlerender.fcgi?tool=pubmed&pubmedid= 11980584

17

Legislation, clinical trials and ethical issues

17.1 Legislative bodies

All therapies destined for use in humans are subject to legislation. Almost every therapy that has been reviewed in the preceding chapters falls under the control of specific legislature. For details of legislation, country-specific websites are available.[287,288,289,290] In the UK, the government has committees that deal with xenotransplantation (UK Xenotransplantation Interim Regulatory Authority), stem cells (UK stem cell initiative), transplantation (Advisory Committee on the Microbiological Safety of Blood and Tissues for Transplantation), gene therapy (Gene therapy Advisory Committee) and vaccines (JCVI/CSM: Committee on the Safety of Medicines/Joint Committee Vaccination and Immunisation Sub-Committee on Adverse Reactions to Vaccines). Indeed, all the information required can be found by searching the Department of Health website.[291] Health EU is responsible for patient safety within the European Union,[292] and their website provides links to legislation concerning many of the therapeutics we have discussed. In the USA, the FDA[293] and NIH[294] are the main bodies responsible for legislation on health- and research-related issues. In other countries, information can be found through accessing government websites.

The National Institute for Health and Clinical Excellence (NICE) (www.nice.org.uk) is responsible for decisions made on availability of drugs and treatments in the NHS in England and Wales. It provides national guidance on promoting good health and preventing and treating ill-health. It is the department frequently in the press for its controversial decisions not to

---

Molecular Therapeutics: 21st-century Medicine by Pamela Greenwell and Michelle McCulley.
© 2007 John Wiley & Sons, Ltd

prescribe drugs, one recent example being Avastin™ for bowel cancer.[295] NICE has the unenviable task of defining cost:benefit ratios and determining the value of drugs in the treatment of patients within the National Health Service. Scotland and Northern Ireland have separate organisations to make decisions. In Scotland NHS Quality Improvement for Scotland carries out appraisals, the Scottish Intercollegiate Guidelines Network (SIGN) provides clinical guidelines and the Scottish Medical Consortium (SMC) gives advice on new medicines. In Northern Ireland the Department of Health, Social Services and Public Safety assesses NICE guidelines to determine whether they are appropriate.

The regulation of the quality and effectiveness of drugs is paramount in public health. There are laws that regulate promotion and advertising that try to ensure that it should be fair and balanced and not encourage drug abuse. Although the regulatory processes are strict, they should not prevent access to medicines.

Regulations ensure that prescribed medicines are of the required quality, safety and efficacy and that health professionals and patients have the information necessary to enable them to use medicines correctly. Of prime importance is the assurance that medicines are manufactured under regulated guidelines, adequately tested for activity, function and toxicity, stored in the optimum environment, and distributed and dispensed appropriately. It is clearly important that there be worldwide sanctions against illegal medicine manufacturing and trade. There is a particular problem when drugs become 'generics', post-patent, when there is a slackening in control of manufacture and the potential for lower-quality materials reaching the market. This is a concern with respect to 'protein drugs' where, for example, incorrect glycosylation will lead to non-functional therapeutic agents. Thus, as these copies of recombinant protein products, termed 'biosimilars' or 'follow-on biologics', will not be identical to the marketed original drug, novel legislation is required before their introduction to the market. Using guidelines set down by the European Medicines Evaluation Agency, biosimilar epoetin products were investigated. The studies reported that products did not have the same composition and showed unacceptable batch-to-batch variation. However, several clinical studies suggested that they did have therapeutic activity. Nevertheless, the analytical results suggest that much more rigorous testing is required with recombinant therapeutic products.[296,297]

There is an excellent review of UK policy on new therapeutics published by the Academy of Medical Sciences UK which addresses many of the current issues, concerns and regulations.[298] This report covers vaccines, gene therapies and clinical trials.

Individual governments are responsible for their own drug regulatory authorities (DRAs) which are accountable to both the government and the public. It is important that these DRAs act in a responsible manner and that their actions are transparent and evaluated. The WHO produces guidelines for 'strengthening national regulatory authorities'.[299] The details are available via government websites.

17.2 Clinical trials

It is obvious that every new medicine destined for use in humans must ultimately be tested in humans. Of course, we do not take a novel therapy and trial it directly on patients or volunteers.

Table 17.1   Clinical trials: outcomes

Type of trial	Outcome
In vitro tests	Can define biological activity in a controlled test, e.g. the effect of recombinant Factor VIII on clotting time.
	Does not involve cells, looks at a single event which may *in vivo* be linked to a range of other events, e.g. clotting cascade.
Cell culture	Can use stem cells or cell lines. Can view toxicity and effect, e.g. *CFTR* gene therapy showed ion pump restoration.
	Does not tell us about the effects of the immune system. Cell lines are often cancer or transformed cells with unusual metabolism.
	Can we trust the results? If we use stem cells, culturing is much more difficult and costly.
Animal studies	Allow us to view metabolism, clearance and see therapeutic effect, e.g. *Factor IX* gene therapy in dogs.
	May not predict outcome in humans due to differences in the immune system and also glycosylation.
Phase I trials	Small numbers of individuals used to test for safety, metabolism and clearance. Choice of volunteer or patient is dependent on nature of therapy.
	For some therapeutics we cannot test on healthy individuals. Testing on very sick patients may not give us useful information.
Phase II trials	Larger numbers of people; for many novel therapies patients who are very sick may benefit from the drug.
	Drugs may fail at this stage due to lack of benefit or because they are not better than currently available drugs.
Phase III trials	A much larger trial with patients who are less sick. It is at this stage we should see benefit.
	Often the point at which drugs are withdrawn, following huge investment due to side effects, lack or efficacy or low cost:benefit ratios.

Some drugs are tested '*in silico*' prior to trials on cultured cells. This would involve, for example, molecular modelling of a drug with its target looking at calculated binding parameters. All potential therapeutic agents are tested in tissue culture to determine whether they are toxic. However, it may be difficult to determine from cell culture trials whether they are effective. Cells in culture may take up drugs, DNA and proteins rapidly as they are bathed in the therapeutic agent. Using this model, it is, however, not possible to determine whether the same drug will enter complex organs or tumours *in vivo*, will be metabolised or will be recognised by the host immune system. Thus the next step is the animal model trial. What is the best model? Clearly, one closely related to humans, but in reality most work is done using rats and mice as these are easier to breed, less expensive and less emotive than, for example chimpanzees. Thus, the results here must be viewed with both knowledge and scepticism and translation of results from animal trials to humans must be addressed. For example, a recombinant protein specifically designed to evade the human immune system may well provoke immune reaction in a mouse. A protein containing the xenoreactive Galα1-3Gal will be accepted by a mouse but not a human. Thus human trials must be carried out, provided there is no evidence of problems in the cell culture or animal trials.

However, for many of the drugs in which we interested, it would be unsuitable to use healthy volunteers. For example, you would not choose to give a volunteer a radioactively labelled antibody or a pig heart transplant. So, frequently volunteers in this type of trial would be patients with the disease under investigation. This poses the following problems:

- The patient may be so sick that any therapeutic effect would be marginal, although in phase 1 trials we are not expecting therapeutic benefit as this really is looking at toxicity, safety, metabolism and clearance.

- The patient may have organ failure which affects drug metabolism; this may affect metabolism of the drug or increase the potential drug toxicity.

- The patient will have been treated with other regimens; do we continue their therapy and will this confound our results?

- The patient may die during the trial of their underlying disease; will we be able to distinguish those dying through drug effects from those dying due to underlying disease?

- The patient may volunteer simply in the hope that their life will be saved; has the patient been coerced by promises of therapeutic action?

Hence there is legislation that controls clinical trials and lays down criteria for patient selection, monitoring and reporting. However, legislation is not worldwide and varies from country to country, leading some pharmaceutical companies to cut corners by trialling drugs in countries where legislation is slack. In theory this is not legal, as there is a worldwide legislation governing research on humans. The 'Helsinki declaration' was first adopted in 1964 and has been updated regularly thereafter. The current version (2004)[300] makes it clear that after a trial the medicine developed should at least be available to those involved in those trials and that drugs should be tested on cohorts from countries likely to benefit from their development. The declaration highlights the adoption of ethical standards that promote respect for all human beings and protect their health and rights. It also recognises that some populations are vulnerable, economically and/or medically disadvantaged and need special protection when being considered as a research cohort. Groups considered to be at risk are those:

- who cannot give or refuse consent for themselves, for example children, the mentally ill and prisoners;

- who may be subject to giving consent under duress, for example prisoners, armed forces personnel;

- who will not benefit personally from the research, for example those in developing countries;

- for whom the research is linked with care or the quality of care, for example those in countries in which healthcare is not normally freely available.

Interestingly, the document also makes the point that experimentation involving human subjects must be carried out to further medical science and states that *'In medical research on human subjects, considerations related to the well-being of the human subject should take precedence over the interests of science and society.'* It is also useful to quote the declaration's standpoint on the purpose of medical research, which it states to be *'research involving human subjects is to improve prophylactic, diagnostic and therapeutic procedures and the understanding of the aetiology and pathogenesis of disease. Even the best proven prophylactic, diagnostic, and therapeutic methods must continuously be challenged through research for their effectiveness, efficiency, accessibility and quality.*' The document also highlights the fact that trialling and use of medicines and therapeutic procedures invoke risks. Finally, it raises the question of ethics in different countries, making it clear that researchers should be aware of the ethical, legal and regulatory requirements in their own countries as well as internationally. The declaration tries to ensure that vulnerable patients and communities who may not understand the implications or ramifications of a trial are not exploited.

You may think that this provides an adequate safeguard worldwide, but the declaration really has little deterrent effect. The declaration makes it clear that trials on human subjects should only be undertaken when there has been an adequate assessment of risk and risk-management. Additionally, if during a trial, the risks are shown to outweigh the potential benefits or there is proof of positive and beneficial results, then trials should be stopped and reassessed. It is obvious that research involving humans should only be conducted if the benefit outweighs the risks to the subject, particularly if the subjects are healthy volunteers. Clinical trials can only be justified if there is a possibility that the trial population will benefit from the results of the research.

17.3 The problems of placebo controlled trials

One of the most interesting statements is: '*The benefits, risks, burdens and effectiveness of a new method should be tested against those of the best current prophylactic, diagnostic, and therapeutic methods. This does not exclude the use of placebo, or no treatment, in studies where no proven prophylactic, diagnostic or therapeutic method exists.*' There has been huge debate about whether 'the best treatment' refers to the best treatment worldwide or locally – this makes a huge difference in the treatment of cohorts.[301]

In the literature there are references to two different protocols for testing new drugs. The First Best Method (FBM) treats one set of patients with the new drug and the other with a known drug (a benchmark), rather than a placebo. The benchmark is the standard treatment in use. This ensures that all the test subjects will be treated. In the Second Best Method (SBM) one half of the test group will receive the drug on trial and the other half a placebo. Which is better?[302] Clearly they are asking different questions, the first can inform researchers how effective the new drug is compared to the current drug whereas the latter asks whether the new drug is better than nothing. The biggest ethical concern is leaving patients untreated and allowing their disease to progress.

However, the biggest potential abuse is when carrying out trials in developing countries, Many companies have worked on the principal that if there is no treatment locally available then a placebo controlled trial is fine. This saves money inasmuch as the company does not have to fund the current best treatment. On the other hand, if we use the FBM, and the best marketed drug is not available in a country, what will happen to those subjects after the trial? Would the companies be responsible for funding continued treatment? For example, AZT is not available in parts of Africa; researchers have tried to investigate whether smaller and therefore cheaper doses were effective in combating HIV. However, there is controversy as to whether the control groups be given the therapeutic dose of AZT or no AZT, which is a placebo; the former is the worldwide interpretation, the latter is the local. The newest version of the declaration takes on board this dilemma and states: 'extreme care must be taken in making use of a placebo-controlled trial and that in general this methodology should only be used in the absence of existing proven therapy'.

However, a placebo-controlled trial could be ethically acceptable when it is necessary to determine the efficacy or safety of a prophylactic, diagnostic or therapeutic method or when the method being investigated is for a minor condition and the patients who receive no treatment (placebo) will not be at risk of serious or irreversible harm. Whether the declaration is effective is questionable and there is clearly room for interpretation of the legislation. Placebo controlled trials are, however, important, with many papers stating that 'the placebo effect' has a role to play in helping or even curing patients.[303,304]

How far can we take this approach? In some cases, patients in surgery trials are exposed to 'sham surgery' in order to ensure that they cannot 'guess' whether they received treatment and then subconsciously influence trial results.[305,306] However, as surgery itself carries a risk to the patient can we ever be justified in treating patients in this way? Likewise, if we are investigating cancer progression is it ethical to give only a placebo to one of the trial groups? Of real interest is the comparison between no treatment and placebo. In an ideal world perhaps we should have three treatment modalities: no treatment, placebo and drug. Each group would provide valuable information; however, from an ethics standpoint we need to be careful that we are not putting the lives of patients at risk.[307]

17.4 The need for informed consent

Although informed consent is mandatory for clinical trial participants, in some countries it has been suggested that consent has been taken simply from a village elder on behalf of his people. There are specific issues of trialling drugs in communities where education or language barriers may lead to individuals taking part in trials that could damage their long-term health.

The key word is 'informed' and thus the consent form should be written in a language that is readily understood by those without medical or scientific backgrounds. Thus, any jargon and technical terms should be avoided. In the UK, NHS organisations are accountable to the clinical governance system for *continuously improving the quality of their services and safeguarding high standards of care, by creating an environment in which clinical excellence will flourish*.[308] The Department of Health in the UK publishes information in a number of languages to address the consent issues and at different levels: for children, parents and doctors.[309,310] Those wishing to research on human subjects, body parts, cells or fluids in the UK must submit an application to their local Ethics Committee for scrutiny.

Within the NHS there is a centralised process and the ethics form that is available from the Central Office for Research Ethics Committees (COREC)[311] is scrutinised for any work involving those treated within the NHS system. Ethics Committees comprise scientists, medical doctors and lay members and it is their remit to determine whether there are any problems associated with the trial. We should note here that in the UK research proposals involving NHS patients must be signed off by the hospital R&D department before work can commence; at this stage there should be due diligence to ensure the scientific rigour of the work.

In the USA (Federal Policy 46.116–111), the following is required as part of the consent form:

- a statement that this is a research project

- an explanation of the purpose of the research and the expected duration of the subject's participation (including an estimate of the total amount of the subject's time involved in participation)

- a description of the procedures to be followed, identification of any procedures which are experimental

- the reason for the subject's selection.

It is also important describe any foreseeable risks or side effects to the subject.

On any consent form, there should be a description of any benefits to the subject or others that may reasonably be expected from the research. Additionally, if there is no benefit to participants this must be declared. To enable a free choice, it is important that physicians disclose alternative procedures or courses of treatment available to the subject, and if the trial involves a change in the standard treatment the subject must be informed.

From the patients' perspective it is important that they understand that they have volunteered for the trial and are free to refuse treatment and leave at any stage. On the other hand, patients must also be aware that the physician may withdraw them from the trial as a result of clinical evidence. Confidentiality is also paramount and the use and storage of data and samples must be discussed with each patient.

If the trial subjects are to be paid then this must also be made explicit on the form. Additionally, if there is the possibility of injury as a result of the research, information as to the medical treatment and compensation available should be included. Of course we might imagine that if a company held the trial, they might encourage the subjects to sign paperwork that would limit any liability if the drugs or procedures prove harmful to the patient. Interestingly, following the TGN1412 drug trial (see below) the clinical trials units in the UK were inundated with calls from potential volunteers who had not realised that they could be paid for participation. Given the circumstances, with six men seriously ill, we might have expected fewer volunteers, but clearly the lure of cash makes people much more likely to take part in trials, regardless of risk.

Despite legislation many very sick patients with no hope of cure using current drugs, may misunderstand the information given when offered a place in a clinical trial. Indeed, one can hardly blame patients who believe they have been selected because their doctor thinks this will help them. It is important to ensure that such patients do not undertake trials without thorough briefing and understanding of potential risks and benefits.

17.5 Trials in developing countries

We have already touched on problems of informed consent and placebo-controlled trials in this context. On a positive note, a laudable development has been the Europe Developing Countries Clinical Trials Partnership

(EDCTP) whose remit is to support a long-term partnership between Europe and developing countries by providing €200 million for the development of new medicines and vaccines against HIV/AIDS, malaria, tuberculosis (TB) and poverty-related diseases. One key aim is to adapt to the needs and the local conditions of the developing countries. This partnership includes EU member states, Norway, developing countries, industry and other financial donors.

The EDCTP programme foresees:

- Management jointly by European and African researchers

- Global advocacy

- Support and training for local researchers and clinicians

- Human resources to guarantee the long-term sustainability of the programme.

The core mandate is the '*support for prioritised clinical trials of new drugs and vaccine candidates that urgently await clinical testing. Such large-scale trials will be conducted in the disease-endemic countries under local clinical, ethical and social conditions, in order to obtain relevant results that directly benefit the populations most in need.*'[312]

There are efforts under way to strengthen ethical review throughout the world through the Strategic Initiative for Developing Capacity in Ethical Review (SIDCER). This is an international project to develop the ethical review of biomedical research globally which has involved the WHO. In October 2002 WHO opened its own ethics unit with the remit of '*harmonizing ethical review standards for clinical trials and ensuring such studies are effectively regulated in WHO's 192 Member States.*'[313]

Clearly, pharmaceutical companies are also realising the public relations value of declaring their position with respect to trials in developing countries. An example is GlaxoSmithKline whose website hosts a document on their standpoint:[314]

GSK-sponsored clinical trials world-wide are conducted according to the same fundamental ethical principles. The studies meet international and national regulatory and legislative requirements and follow the research methodologies outlined in the International Conference on Harmonisation (ICH) Good Clinical Practice guidelines.[315] Moreover, GSK-sponsored clinical trials follow the principles contained in the World Medical Association Declaration of Helsinki on the Ethical Principles for Medical Research Involving Human Subjects.

Clinical trials are only conducted in countries where the medicines are likely to be suitable for the wider community. Furthermore, clinical trials of investigational medicines are not conducted in countries when it is known at the outset that there is no intent to pursue registration and make the medicine available for use in that country.

Similar sentiments can be found on websites of other drug companies.[316] Clearly, we could dedicate a whole book to legislation and ethics and another to the ethics of trials in developing countries. The following links will take those of you who are interested onto sites where the ethical problems are discussed in more detail.[317,318,319,320,321,322]

17.6 Recent trial issues

In the USA, the case of Jesse Gelsinger brought into question whether rules were being policed. As you may recall, Jesse was involved in the clinical trials of adenoviral-based gene therapy. Although he was controlling his disease (ornithine transcarbamylase deficiency) with drugs, he was recruited onto the trial and given the highest dose of vector. Days later he died, following a reaction to the therapy. His parents claimed that he was not informed of the risks and that he had not been told of side effects seen in animal models. In terms of ethics, we need to wonder about the ethical committee that sanctioned trials on people whose disease was controlled, particularly given that gene therapy is experimental.

Following his death, there was an investigation to determine the cause and to minimise the chances of further deaths. The Food and Drug Administration and the National Institutes of Health launched an inquiry and, to their dismay, discovered that researchers were not following the federal rules that required the reporting of any adverse events associated with the trials. Indeed, some scientists were reluctant to make problems public, as this might affect future funding. More worryingly, allegations were made about six other unreported deaths attributed to genetic treatments. It was suggested that only 35 to 37 of 970 serious adverse events seen in gene therapy trials were ever reported to the NIH. Since then, there has been tightening of legislation, policing and reporting, but it is clear that it is difficult to ensure that scientists do not breach legislation.[323]

In the UK, we have recently seen an example of serious side effects of a drug in clinical trials. During a phase I trial (the first time the drug was tested on human subjects) of a new drug, TGN1412 (an immunomodulatory humanised monoclonal antibody intended to treat chronic inflammatory conditions and leukaemia), six men became seriously ill and were hospitalised in intensive care. Two were in a critical state with organ failure. Subsequently, there have been reports of the development of precancerous growths in some of the participants. This trial had involved eight men, two of whom were no-treatment controls. The UK Medicines and Healthcare Products Regulatory Agency (MHRA) stopped the trial pending investigations.[324]

The adverse effects associated with the TGN1412 trial is an example where a drug worked well in animal models, but life-threatening side effects and no therapeutic benefit were seen in humans. New safety measures specifically for therapeutic monoclonal antibodies have been proposed as a consequence of this trial, as induction of aberrant immune responses can be catastrophic.[325] The European clinical trials database (EUDRACT Database) has been established to encourage the sharing of relevant information between scientific authorities.[326] This should enable cross-checking of trials data to ensure that problems are swiftly communicated. The website also has detailed guidance for those requesting authorisation for a clinical trial on a medicinal product in the EU states.

17.7 Conclusion

The health of the population of a country impacts on the health of the economy. There have been improvements in legislation for clinical trials over the last two decades that have emerged because of a growing concern that humans should not be treated simply as guinea pigs. The parity of treatment and care is paramount, but in reality, in developing countries the healthcare budgets, as low as $10 per person per year, cannot cope with increased demand for modern drugs. Some pharmaceutical companies have offered free or low-cost drugs to developing countries. This has worked well in some cases but in others the drugs have been sold on by unscrupulous governments. Some countries have refused aid, fearful that they will be asked to pay in kind. We need to ensure that legislation protects vulnerable groups from exploitation, and that drugs should be tested on communities with diseases that will benefit from the drugs being trialled and that their governments can actually afford them. There is no excuse for having double standards.

Points to consider

If you were NICE and had £100 million to spend, but the amount required to satisfy the needs of all the patients was £200 million, what criteria would you use to decide which drugs to supply and which patients to treat?

How can we ensure that developing countries and, more importantly, the sick within those countries, benefit from the newly developed drugs?

Notes

287 www.parliament.uk/
288 www.fda.gov/
289 www.europarl.europa.eu/
290 www.opsi.gov.uk/legislation/index.htm
291 www.advisorybodies.doh.gov.uk
292 http://ec.europa.eu/health-eu/care_for_me/patient_safety/index_en.htm
293 www.fda.gov/
294 www.nih.gov/
295 www.nice.org.uk/aboutNICE
296 Combe, C. et al. Biosimilar epoetins: an analysis based on recently imple-
     mented European medicines evaluation agency guidelines on comparability
     of biopharmaceutical proteins. Pharmacotherapy, 2005, 25, 954–62
297 Crommelin, D. et al. Biosimilars, generic versions of the first generation of
     therapeutic proteins: do they exist? Contrib. Nephrol, 2005, 149, 287–94
298 www.acmedsci.ac.uk/images/page/1132655880.pdf
299 www.who.int/medicines/areas/quality_safety/regulation_legislation/en/
300 www.wma.net/e/policy/b3.htm
301 Cox, P. Codes of medical ethics and the exportation of less-than-standard
     care. Int. J. Appl. Philos., 1999, 13(2), 177–85
302 www.bu.edu/wcp/Papers/Bioe/BioeCool.htm
303 Benedetti, F., Rainero, I. and Pollo, A. New insights into placebo analgesia.
     Curr. Opin. Anaesthesiol., 2003, 16, 515–19
304 Miller, F.G. and Rosenstein, D.L. The nature and power of the placebo effect.
     J. Clin. Epidemiol., 2006, 59, 331–5
305 Wolf, B.R. and Buckwalter, J.A. Randomized surgical trials and 'sham'
     surgery: relevance to modern orthopaedics and minimally invasive surgery.
     Iowa Orthop. J., 2006, 26, 107–11
306 Heckerling, P. S. Placebo surgery research: a blinding imperative. J. Clin.
     Epidemiol., 2006, 59, 876–80
307 http://content.nejm.org/cgi/content/short/344/21/1594

308 www.dh.gov.uk/PolicyAndGuidance/HealthAndSocialCareTopics/
     ClinicalGovernance/fs/en

309 www.dh.gov.uk/PolicyAndGuidance/HealthAndSocialCareTopics/Consent/
     Consent GeneralInformation/fs/en

310 www.dh.gov.uk/assetRoot/04/06/69/93/04066993.pdf

311 www.corec.org.uk/

312 http://ec.europa.eu/research/info/conferences/edctp/pdf/edctp-at-a-glance_en.
     pdf

313 www.who.int/bulletin/volumes/82/4/ethics0404/en/index1.html

314 www.gsk.com/responsibility/Downloads/clinical_trials_in_the_developing_
     world.pdf

315 The International Conference on Harmonisation of Technical Requirements
     for Registration of Pharmaceuticals for Human Use (ICH) is a unique project
     that brings together the regulatory authorities of Europe, Japan and the
     United States and experts from the pharmaceutical industry in the three
     regions to discuss scientific and technical aspects of product registration.
     www.ich.org/cache/compo/276-254-1.html

316 www.roche.com/pages/facets/18/ethicstriale.htm

317 www.bioethics.nih.gov/international/declarat/avis17_en.pdf

318 www.pubmedcentral.nih.gov/articlerender.fcgi?artid=1183235

319 www.pubmedcentral.nih.gov/articlerender.fcgi?artid=1126734

320 www.bioethics.nih.gov/international/readings.html

321 www.nuffieldbioethics.org/go/ourwork/developingcountries/introduction

322 www.nhmrc.gov.au/ethics/human/issues/trials.htm

323 www.fda.gov/Fdac/features/2000/500_gene.html

324 www.newscientist.com/article.ns?id=dn8852

325 Liedert, B. *et al.* Safety of phase I clinical trials with monoclonal antibodies
     in Germany – the regulatory requirements viewed in the aftermath of the
     TGN1412 disaster. *Int. J. Clin. Pharmacol. Ther.*, 2007, 45, 1–9

326 http://eudract.emea.eu.int/

Epilogue

The authors hope that we have stimulated your interest in the subject area and also heightened your awareness of the problems of developing novel therapeutics. We would also like to imagine that we have made you think, particularly about the ethical problems and made you more critical and less accepting about material you read. We have introduced you to websites that are generally well informed and, on the whole, readily available to you. Use them, stay informed and help to inform others.

Sourcing references

References from the internet change rapidly and we are mindful that you may find that some references will have moved. We therefore recommend the following.

The NCBI website www.ncbi.nlm.nih.gov can be used to access Pubmed, PubmedCentral (free full text, online books, OMIM (Online Mendelian Inheritance in Man) – a superb information resource for inherited diseases). Searching can be frustrating, but using the LIMITS button in PubMed, you can limit the returned articles to full text and ask for reviews. Search engines such as www.google.com (or Google Scholars) or www.alltheweb.com allow you to search for government websites, companies, news, images, videos and audios. Be careful to ensure that the sites you choose are bone fide. University, company and government websites are good for information.

The FDA, WHO, UK parliament, Department of Health (UK) are all simple to find using these tools. However, the authors would be delighted to help you find more information on specific topics: please feel free to email on greenwp@wmin.ac.uk

Molecular Therapeutics: 21st-century Medicine by Pamela Greenwell and Michelle McCulley.
© 2007 John Wiley & Sons, Ltd

Index

α-1-antitrypsin 94
A antigen 3, 33, 34
abortion 5, 12, 13, 107, 143
active immunity 45
adenoassociated viruses 152, 154, 177
adenosine deaminase deficiency 100,
    149, 163, 165,
adenoviruses 151, 171, 212
affinity purification 26
AIDS 2, 17, 51, 57, 77, 190, 227, 240
AIDSVAX 57
Alder Hey 142
allograft 69, 99, 104, 139
Alzheimer's 13, 16, 159, 195
animal models 1, 2, 5, 7, 8, 23, 38, 94,
    143, 145, 152, 171, 181, 197, 203,
    211, 241
Anthony Nolan Trust 102
antigen presenting cells 46
anti-RhD 67
anti-sense 181, 211
autism 49
autograft 69, 99
Avastin 232
AZT 51, 237

B cells 46, 66, 69, 74
Battens disease 100, 157
BCG 47, 49
bcr-abl 197, 211, 218, 220
Behring 44
beta-galactosidase 26, 174

Bill and Melinda Gates Foundation 53
blastula 87, 125, 138
bone 40, 103, 106
bone marrow 8, 69, 77, 99, 109, 130,
    137, 140, 157, 166, 178, 197, 209,
    211
Borna virus 39
brain 107, 109, 120, 140, 144, 158,
    174, 186, 213
BSE 84, 86, 118
bystander effect 163, 171, 215

cancer 12, 33, 39, 60, 63, 65, 131,
    151, 159, 166, 169, 178, 181, 187,
    205, 223, 232, 233, 242
Cantab 70
CCR5 3, 227
CD20 74
CD37 74
CD45 70
cell cycle 4, 126, 127, 181, 206, 207,
    212
CF 12, 15, 104, 163, 170, 178
CFTR 7, 84, 96, 163, 170, 178, 233
chemotherapy 4, 8, 72, 77, 79, 101,
    106, 209, 217
chimeric cloned blastulas 142
Chinese hamster ovary cells 35
CJD 18, 37, 39, 107, 110, 118, 159
clinical trials 1, 56, 70, 76, 119, 145,
    173, 175, 189, 197, 202, 213, 216,
    219, 220, 228, 231

Molecular Therapeutics: 21st-century Medicine by Pamela Greenwell and Michelle McCulley.
© 2007 John Wiley & Sons, Ltd